AFE96242

Neuroleadership

James Teboul · Philippe Damier

Neuroleadership

Creative Leadership with a Focus on the Brain

Springer

James Teboul
Emeritus Professor
INSEAD
Fontainebleau, France

Philippe Damier
Professor of Neurology
Nantes University
Nantes, France

ISBN 978-981-99-5121-5 ISBN 978-981-99-5122-2 (eBook)
https://doi.org/10.1007/978-981-99-5122-2

© The Editor(s) (if applicable) and The Author(s), under exclusive license to Springer Nature Singapore Pte Ltd. 2023

This work is subject to copyright. All rights are solely and exclusively licensed by the Publisher, whether the whole or part of the material is concerned, specifically the rights of reprinting, reuse of illustrations, recitation, broadcasting, reproduction on microfilms or in any other physical way, and transmission or information storage and retrieval, electronic adaptation, computer software, or by similar or dissimilar methodology now known or hereafter developed.
The use of general descriptive names, registered names, trademarks, service marks, etc. in this publication does not imply, even in the absence of a specific statement, that such names are exempt from the relevant protective laws and regulations and therefore free for general use.
The publisher, the authors, and the editors are safe to assume that the advice and information in this book are believed to be true and accurate at the date of publication. Neither the publisher nor the authors or the editors give a warranty, expressed or implied, with respect to the material contained herein or for any errors or omissions that may have been made. The publisher remains neutral with regard to jurisdictional claims in published maps and institutional affiliations.

This Springer imprint is published by the registered company Springer Nature Singapore Pte Ltd.
The registered company address is: 152 Beach Road, #21-01/04 Gateway East, Singapore 189721, Singapore

Paper in this product is recyclable.

Acknowledgments

It's time to move on to a realistic approach to leadership by referring to what we already know about the brain.

We hope to contribute to this by opening opportunities for seminars on the subject and developing networks of people seeking to take their organizations forward by enabling their members to understand themselves better, and develop a context and a culture that nurture solidarity and trust.

We must thank all the authors cited in the bibliography who contribute to the study of this field and whose work has fed into this book. We drew additional inspiration from David Rock, the director of the NeuroLeadership Institute, who applies what is known in neuroscience to the practice of management and the development of leaders.

Our thanks to INSEAD which has supported us, and in particular professors Steve Chick, Manuel Sosa for the joy of much discussion, as well as Enrico Diecidue for introducing some notions of neuroleadership in his teaching.

Our thanks to Margaret Morrison for her excellent help with adapting, reviewing, and editing the book. Our thanks also to Roger Pring, a talented illustrator whose invaluable help has already contributed greatly to our previous work.

We should not forget to thank our students for their candid questions. They may make us mutter under our breaths because they often take us out of the framing we have in mind, and we may well not know how to respond to them immediately. But we are committed to preparing an answer for the next lesson. We are learning.

Our thanks to Philippe Mahrer, the director of the Collège des Ingénieurs, always interested in new pedagogical approaches.

We would also like to thank Anushangi Weerakoon at Springer, who has been a great help and provided advice throughout all the stages of the preparation of this book.

Introduction

Leading seems natural and innate. Some people are at the helm, responsible for leading toward a common goal, whether they were given power or managed to acquire it, and others defer to authority and are following. Leading and being followed seems straightforward. This is something that we regularly experience in our history books, in our political or economic environment, or in the many chains of command and pyramidal structures we live in. The problem is that people are less and less satisfied with the way they are led today, in an environment that is increasingly complex, uncertain, and competitive. They are less ready to accept authority and less inclined to follow. It appears more and more clearly that leadership does not exist without followership and cannot be reduced to the exercise of authority.

A Definition

The notion of leadership must be redefined beyond the traditional vertical exercise of power, which is in fact the easy way of leading. Leadership is more demanding and must take into account the *process* that links leaders, followers, and the context in which they are situated. Leadership should engage leaders and followers to collaborate and learn within a motivating cultural context. This is a more effective and creative way of leading, but much harder.

Telling people where to go and expecting to be followed is not leadership. To get people ready to follow, leaders must show a true commitment, mobilize their followers, and organize the required cultural context. The game is more open and more complex, as the engagement of followers can be activated not only by external motivators such as monetary rewards or status, but also from deep motivations such as finding more meaning in what they do, gaining more freedom of action, more social recognition, or inclusion.

The Creative Model

This extended notion of leadership has been developed by scholars and consultants in management education and has resulted in a profusion of publications, papers, and books. It has opened the way to multiple prescriptive models that look quite reasonable and enticing. We have summarized them into a reference model, the *creative model of leadership*, that tells leaders what to do in theory. But that does not help them much in practice given the levels of dissatisfaction, stress, or unhappiness that are generally observed. Organizations tend instead to backslide toward a regressive model that we call the *fixed model of leadership* because of its rigidity and its Taylorian flavor. As a result, leaders navigate between these two models, depending on the efforts they are prepared to make and the discipline they will follow.

So, why is it so difficult to implement a prescriptive model like the creative model? Why is it so difficult to align the motivations of leaders and followers?

Leadership is Quite Demanding

Leaders most frequently fail to follow the prescriptive model because they don't take into consideration the limitations of their own human nature and the difficulty of exercising self-discipline, as well as the importance of motivating followers. The road to the creative model is harder to follow because leaders will not trust their followers even when they empower them, because followers will not collaborate if they do not obtain the freedom, the meaning, and the recognition they expect, or because the supportive culture of trust and exchange does not persist, and so on. Understanding the complexity of relational interaction is quite demanding, and leaders are tempted to take the easier road of command-and-control, expecting that people will align and obey, while using threat or extrinsic rewards. They allow themselves to drift back toward the fixed model approach, which is more comfortable, but is not leadership. To engage in the creative leadership approach, leaders must take into consideration the complexity of their own human nature and more specifically the way their brains function.

Exploring the Capabilities and Limitations of the Brain

Just like sport professionals need a good understanding of the functioning of the body to exercise effectively, leaders need to practice and explore the capabilities and limitations of their brains to be more effective. They should realize the extent to which they depend on neuronal processes operating at a subconscious level. The brain was not initially designed to adapt to the level of relational complexity and the rapid environmental evolution that it is facing today. The brain of Homo Sapiens was mainly constructed to help survive in the savannah and interact within small communities.

Leaders must now realize that rationality is overrated, that emotions emerged before reason, even though emotional intelligence, only documented recently, was considered as a revelation. They must also keep in mind seven capital biases that influence their behavior. Among them, the hierarchical bias invariably takes them back towards centralization and vertical exercise of power. The immediacy bias makes them focus on the short term, the confirmation bias limits their creativity, and so on. Leaders need to be aware of specific predispositions and biases that condition their action and must learn how to use motivators to engage all parties in thinking, collaborating, and learning together, in the same direction.

A Disciplined Process

Depending on the efforts they are prepared to invest, leaders will navigate between the fixed and the creative model. If they are serious about transforming their organization with an effective leadership approach, they must put in place a process that will engage them with discipline on the hard and long road that leads to the creative model. They will take a route which is less spectacular at first sight and more demanding in the long run. They will create an environment where people are motivated to work and learn together and find meaning in what they do. If not, they will let go and fall back on a rigid approach, exercising raw authority on followers who will drag their feet. This is hardly leadership. People can live with it but it's painful.

Leadership is hard work because this approach necessitates the implementation of a disciplined process that gives followers long-term perspectives, fires their passion for a cause, motivates them to master what they do, or gives them a sense of being part of an active community that is going to have an impact.

Contents

Chapter 1 opens with a better understanding of what leadership is about. Leadership cannot be dissociated from followership. Chapter 2 focuses on the traditional rational approach of leadership, exemplified by the Taylorian bureaucratic model. Its effectiveness was not only based on its methodical approach but also on the fact leaders could lead according to their natural inclinations and followers had no other choice but to follow. To go beyond this simplistic approach, it is necessary to find out what prevents leaders from coming up with a more creative model where they can associate and collaborate with their followers toward a common goal. A good reason for this limitation is the absence of an adequate understanding of the predispositions and capabilities of their own brains. So, Chap. 3 starts with a simplified overview of the brain in three dimensions and the way it works, which we might categorize as emotional, and Chap. 4 revisits the notion of rationality and the importance of emotions and emotional intelligence.

We are now ready to analyze how the rigid Taylorian model, which hampers the individual's development at work, has triggered a revolt in employees, co-workers, and customers, and consequently initiated many attempts to develop a more suitable leadership approach (Chap. 5).

All these attempts can be summarized by the model that we call the "creative model", a reasonable model that promises success, innovation, and satisfaction for all. Toyota provides a pioneering example that has successfully been followed at Virginia Mason (Chap. 6).

The creative model of leadership is based on three pillars: a long-term strategic approach (Chap. 7), a motivating leadership style (Chap. 8), and a culture of collaboration and learning (Chap. 9). In each of these chapters, the question remains whether this type of model is applicable over time, or if it is just a mirage that is difficult to achieve. We show how this approach is weakened by our predispositions and biases and thus often remains inaccessible.

We then take an overview of the functioning of the human brain and consider the seven capital biases that have been described in previous chapters. We also outline a method for dealing with these biases in everyday life (Chap. 10).

We go on to formulate the conditions for success before ending with a somewhat extreme example from Netflix (Chap. 11).

Our conclusion establishes a roundup of the current situation and advises that human beings, their cognitive processes, and their aspirations, ought to be repositioned at the heart of the way creative leadership should be practiced within organizations.

N.B.: For ease of reading and maximum clarity, the number of notes throughout the text has been limited. The bibliography cites all the authors to whom this work is indebted.

Contents

1	**Leadership, Followership, and Context**	**1**
	The Great Man	1
	Focusing on the Leadership Process	7
	Leadership of Collaboration and Learning	9
	Exploration and Execution	9
	Importance of Context and Culture	11
	The Creative Leadership Model	12
	The Limit of Prescriptive Models	13
	Neuroscience can Help	14
	References	14
2	**A Straightforward Rational Managerial Model**	**15**
	Industrial Revolution and Mass Production	15
	The Scientific Organization of Work. A Rational Approach	16
	The Fixed Model of Leadership	21
	Reference	22
3	**A Brain in Three Dimensions**	**23**
	Why Do we and Other Animals Have a Brain?	23
	Vertical Integration: The Triune Brain	23
	Left–Right Integration	26
	Back-to-Front Integration	29
	References	34
4	**A Very Emotional Brain**	**35**
	Revisiting the Notion of Rationality	35
	A Useful Brain Valuation System	37
	What is the Point of Emotion?	39
	The Bias of Emotional Reasoning	40
	The Fear Circuit	41
	The Reward Circuit	43
	Self-Control and Willpower	46
	References	48

5 **The Rebellion** 49
Employees Rebel 49
Customers Rebel 51
Toward Transformational Leadership 53
References 57

6 **Creative Leadership in Action** 59
Toyotism 59
A New Strategic Vision 60
An Engaged Leadership Style 61
Shaping the Context 62
Reinforcing the Culture and Orchestrating the Whole 64
Turning Around Virginia Mason 66
The Strategic Vision 67
Leadership Style and Governance 69
Organizing and Redesigning the Context 70
A Fluid Context and a Learning Culture 72
Orchestrating the Whole 73
References 75

7 **Strategic Thinking** 77
Rational Thinking 77
Predictive Intelligence 79
Innovation and Creativity 82
Strategic Thinking is Constrained by Our Biases 84
Reference 87

8 **Leadership Style** 89
The Leader's Personal Commitment 89
Sharing the Vision to Mobilize Followers 93
Mass Empowerment. Everyone as a Potential Leader 94
References 99

9 **Establishing the Right Context and Culture** 101
Hiring the Right People 101
Setting up the Proper Context 103
Practicing a Culture of Relational Engagement 104
What Limits Relational Engagement? 106
Practicing a Learning Culture 108
What Limits the Learning Culture? 110
Orchestrating the Whole 111
What Limits Overall Evolution and Orchestration? 112
References 113

10 Accepting Our Brain as It Is ... 115
Back to the Overall Picture ... 115
Our Brains are Biased ... 118
Overcoming Biases ... 119
Mitigating the Seven Capital Biases ... 120
Bias of Fear and Psychological Insecurity ... 121
Bias of Immediate Reward and Future Discounting ... 122
Egocentric Bias and Overconfidence ... 124
Hierarchical Bias of Domination and Control ... 125
Bias of Inertia and Least Effort ... 126
Confirmation and Framing Bias ... 127
Bias of Similarity and Social Conformity ... 129
A First Step but This is not Enough ... 131
Reference ... 131

11 A Disciplined Process in Action ... 133
A Disciplined Approach to Generate Motivation and Swim Against the Tide ... 133
Sharing the Vision with Determination ... 134
Hiring the Right Talent ... 135
Distributing Power to Facilitate Experimentation ... 136
Setting the Proper Context and Culture ... 137
Rewarding and Orchestrating the Whole ... 140
Leadership According to Jim Collins ... 141
Netflix's Innovative Approach ... 143
References ... 148

12 Conclusion. Keeping the Human Side in Mind ... 149
Two Benchmark Leadership Models ... 149
The Importance of Knowing How the Brain Functions ... 151
Putting Relational Culture Back at the Heart of Things ... 151

Annex 1—A Handbook ... 153

Annex 2—The Big Five Personality Traits ... 157

Annex 3—We Are Our Synapses ... 161

Annex 4—The Fear Circuit ... 167

Annex 5—The Reward System ... 171

Annex 6—The Insula: The Body Sensor ... 175

Annex 7—Habits ... 177

Annex 8—Teaching and Practicing Biases ... 181

Bibliography ... 185

About the Authors

James Teboul has been the director of the International Executive Program at INSEAD for many years. He has taught in many executive programs and has been a consultant to industrial and service organizations. Graduate Engineer from Ecole Centrale Paris and Doctor ès Science from Paris University, James Teboul started his career as a group leader responsible for new product development at Schlumberger. After obtaining an MBA at Sherbrooke University (Canada), he became a professor at INSEAD. He was a member of the editorial board of the International Journal of Service Management and has published several books in the domain of management and leadership.

Philippe Damier is a full professor of Neurology at the University of Nantes. Among his responsibilities at the Medical School, University Hospital of Nantes, he chaired both the Department of Neurology and the Clinical Research Center. He is currently in charge of the expert center in Parkinson's disease for the west part of France. He has published more than 250 scientific articles in international journals, most of them in the field of Parkinson's disease and movement disorders. In 2011, he got a MBA in Melbourne (Australia). He is the author of '*Decider en toute connaissance de soi*', O. Jacob.

Chapter 1
Leadership, Followership, and Context

Organizations need leadership to remain vibrant and adaptive in a world in constant and rapid evolution. They need leaders to show the way forward and convince groups of individuals, who may not be well-known to each other, to follow them and cooperate to achieve a common goal. This capacity allowed puny Homo sapiens to dominate his environment. Alone he is powerless and amounts to almost nothing, with cooperation he has unlimited power. As the environment becomes increasingly volatile and complex, stress and anxiety are on the rise, and enlightened leaders are needed to reassure and provide meaning.

Leadership has thus generated a tremendous interest as can be seen in the thousands of publications dealing with the subject, but its definition remains elusive and has evolved over time. The focus has shifted from the leader as the Great Man in charge from his lofty helm to leadership as a process that deals with the way leaders and followers interact in a specific context.

The Great Man

The first theories to emerge put the 'Great Man' at the top, showing the way ahead. For example, the British historian Thomas Carlyle[1] asserted that history is dominated by the deeds of great men. The leader is then represented by heroic or demonic figures, military commanders and generals, heads of state, kings or emperors, often dictators or despots, who are supposed to exercise a huge influence on the course of history. They were able to lead great numbers of people in often audacious directions into which followers would never have otherwise ventured. The leader, viewed as an intrepid navigator in full control, never seriously questioned the willingness of followers to follow, nor considered the complex nature of the leadership process. Leadership was equated with the power that leaders could wield in order to be

[1] Carlyle (1841)

© The Author(s), under exclusive license to Springer Nature Singapore Pte Ltd. 2023
J. Teboul and P. Damier, *Neuroleadership*,
https://doi.org/10.1007/978-981-99-5122-2_1

followed, and most often, they only were able to show their influence in crisis or wartime.

Domination and Control

In fact, the easiest way to ensure followership is to exercise power vertically, top down, by using rewards or applying physical control and threats, in short by using a combination of carrots and sticks to get people aligning in the right direction.

This command-and-control approach is in fact quite comfortable as leaders can enforce their vision and dictate the actions of their followers unilaterally. But are they effective leaders?

Hierarchical Bias

The exercise of vertical power operates according to a *hierarchical bias* that is engraved in neural circuits in our brains. Since the dawn of time, hierarchy has imposed itself to allow increasingly large groups of individuals to function together. This primary form of organization is found in many animal species. For example, chimpanzees live in groups of individuals with a dominant animal, the alpha male, to whom the other members clearly signal their submission. The dominant male strives to bring social harmony to his group, not only by force, but by leading a larger and more stable coalition of supporters than his competitors. Political power, in short.

As soon as human groups became sizeable, especially after the emergence of agriculture, hierarchies-often very rigid ones-were established with leaders amassing more and more power. This power was exercised on people readily prepared to accept authority. This 'voluntary servitude' was pointed out by the French writer Etienne de La Boetie who questioned the fact that tyrants remained in power and people willingly submitted to their rule when it was against their interest to do so. Power does not only rely on violence or coercion, but on the people's willingness to obey their leaders and even attempt to please them. This tendency to willingly follow orders has been further illustrated by Stanley Milgram's famous experiment on obedience to authority[2].

Participants in the experiment, men from a diverse range of occupations and varying levels of education, were instructed to administer electric shocks to a hidden "learner" if he failed to remember the words he was taught. These fake electric shocks gradually increased to levels that would have been fatal,

[2] Milgram (1974).

had they been real. The experiment found, unexpectedly, that two-thirds of participants fully obeyed the instructions and administered shocks even after the subject of the experiment had become completely unresponsive.

In the same way, people who feel under threat when faced with uncertainty and rapid change willingly accept the authority of a leader who is supposed to guide and reassure them.

Power Based on Fear

Dominant leaders basically utilize levers of reward and punishment to coerce and compel people to do what they want. Power based on fear is particularly effective as it relies on one of our most primitive concerns, a natural bias to avoid threat and insecurity. It's about staying alive and safe.

The Fear and Insecurity Bias

Deeply embedded in our brain is a system which deals with fear. This is the result of our evolutionary history, as threats, in the form of natural hazards or aggressions of all kinds, have had a great bearing on our survival. This circuit is set in motion by the *amygdala, the brain sentinel* which constantly monitors our environment. As soon as an unforeseen event occurs, the amygdala activates and triggers the release of *neuromediators* that prepare our body to flee or fight. As far back as Machiavelli, 500 years ago, the advice to leaders was to rely more on fear and power than on trust or reason.

Power as a Relationship

Power, in fact, is not something you can possess like a crown, a title, or a form of energy. Power manifests itself in the relationship between one person who has specific resources and another who has specific needs. Power is thus the ability to influence others, by controlling the resources desired by those others, such as status, rewards, or information. You only have power over another person if you can dispense the resources that are needed by that person. If I have an urgent need to go to town, and my car is broken, you are able to exercise some power over me by virtue of your ability to lend me your car.

Legitimate Authority

Followership is also obtained through obedience to a legitimate source of power established by tradition, religious endorsement, popular acclaim, or rights of succession. As these leaders assumed certain responsibilities, preserved order, and gave protection in the event of danger, they had the right to command, and their subjects the obligation to obey. Only a few complained, as the fundamental need of the people was for order and security.

We should note here that the exercise of authority is often based on a 'power distance[3]' which depends on the cultural norms in place. A high power distance, seen in many Asian cultures, imposes respect and obedience, with strictly defined ranks in families and societies, while the distance is considerably reduced in more egalitarian Scandinavian cultures.

Fusional Leadership

Other leaders gain status when their inspired vision mobilizes zealous followers who will idealize them. As they identify with their leader, followers become highly dependent and easily manipulated. They are unified by a common political or mystical ideology, and they disappear as autonomous person. Stadium demonstrations in totalitarian countries illustrate this metaphor with the real phenomenon of the crowd holding aloft panels which read together form huge slogans. There is no room for personal thought and woe to anyone who does not follow the groupthink. The Party is always right.

The most extreme cases are demonstrated in people devoted to a cause, passionately committed to "sacred values", whether religious or secular, which transcend socio-economic and power interests. They are willing to sacrifice everything for their cause and will blindly follow the leader who represents it.

Trait Approach of the Great Man

The Great Man approach is based on the idea that leaders are born, not made. 'Natural' leaders have innate qualities that set them apart from others, such as size, intelligence, self-confidence, decisiveness, determination, integrity, or eloquence. A long list of qualities that make us forget that the most successful leader endowed with all these qualities once they are no longer successful, is suddenly nothing but a loser. Nevertheless, some of these qualities are useful and they can be recognized in some of

[3] Hofstede (2010).

the Big Five[4] personality factors (see annex 2). For example, determination and integrity seem to correlate well with the *Conscientiousness* factor, and decisiveness with *assertiveness*.

Napoleon used two dimensions: intelligence and energy; and Jack Welch,[5] the former CEO of General Electric had a compelling mantra of the 4 E's: 'Energy, Energize, Edge, Execute'. In this last case, we recognize determination (energy and energize) and decisiveness (edge and execute).

In fact, these supposed qualities cannot be generalized as we have seen all sorts of personalities at the helm, some seem well balanced, but others can also be suspicious, compulsive, histrionic, or cyclothymic personalities who can be quite influential as long as they can wield power and get results.

However, personalities demonstrating outgoing and dominant behaviors are regularly chosen as leaders. Their self-confidence is often boosted by a narcissistic tendency based on their egocentrism.

Egocentric Bias

The egocentric bias refers to our self-interest, the interest in whatever relates to oneself. As they become more assertive with real power, leaders become attracted to anything that can boost their ego such as professional success, compliments, and applause.

In fact, narcissistic leaders are recruited with the assumption that the person in charge of an organization, a company, or a project should have an outgoing and dominant style in order to change the destiny of the unit. This often leads to organizations headed by narcissistic bosses who become overconfident and give more credit to their own beliefs and creations than to similar ideas generated by others. This excess of self-confidence limits their ability to truly listen and accept feedback and evaluations from others. Instead, they prefer to indoctrinate and dominate. We might note that these behaviors are typically more masculine than feminine and could be a hindrance to the career of women or, on the contrary, offer them more opportunities when the power is more evenly distributed.

From Innate Qualities to Competencies that can be Acquired

As it was difficult to produce a well-defined list of leadership qualities to rely on, researchers and scholars focused on a repertoire of *competencies* that could be developed over time to prepare to effective leadership.

---

[4] This psychological model aimed at describing personality was designed by Lewis Goldberg, then developed by Paul Costa and Robert McCrae in the 1980s.

[5] The outspoken former CEO of General Electric achieved fantastic results in the short term, but more debatable ones in the longer run.

Competencies such as personal effectiveness and better communication skills, self-confidence and delegation style, teamwork and conflict resolution can be learned and improved. The development of these skills is open to all through field practice and training programs. However, these competencies remain rather general, and innate personality traits often overlap with competencies.

Too Much Attention to the Leader

Our brain is naturally predisposed to explain complex situations by simplifying them and by focusing on specific events and the actions of a few outstanding actors. As will be shown in Chap. 3, the processing of conscious information in real time is quite restricted given the limited capacity of working memory available in the brain. Like a tragedy in three acts or a good novel, a compelling story narrates the remarkable actions of few key actors to whom we attribute exaggerated capabilities and authority. Moreover, a coherent scenario is more easily memorized and is used as a template to tell a heroic story.

Influence of the Context

By assigning too much importance to the actions of some leaders over the short term, the risk is to ignore the influence of the context and the culture in which all the actors operate. And this influence can prove decisive. Among historians, there is an unresolved debate as to whether history's course depends more on specific circumstances and context than on particular leaders. Some argue that leaders appear to be great and influential not because of their personal qualities but because they pursue policies resonating with views already held by their followers, or because they take advantage of the opportunities that they enjoy at the time (crisis or war). For example, Lincoln's presidency took place during the American Civil War, and Churchill, Roosevelt, and de Gaulle had their most decisive influence in wartime or during the Algerian uprising[6].

Fernand Braudel, one of the greatest modern historians, used to make the distinction between short and long term. The history of heroes written in the short-term often ignores the long-term influence of the context and the prevalent culture. He advocated that we shift our attention from the analysis and manipulation of the history of individuals or events and consider the evolution of larger systems over time through the effects of geography, climate, technology, or human culture.

Up to now we have trodden an *easy path* as we looked at a leader managing top down, focusing on the *"what-to-do"* and gaining effective results. We now must

[6] Diamond (2019)

take *a hard path* with some humility to deal with the process of leadership, the *"how-to-do"*, how to get followers on board and organize the context.

Focusing on the Leadership Process

The vertical leadership approach is straightforward and comfortable because it is easy to lead when people do what they are told. But this approach is no longer sustainable, as the world keeps changing rapidly, organizations get more complex, and people demand more power sharing and participation. Leaders also need to keep an eye on the way they interact with their followers. Leadership does not happen in isolation, as there can be no leader without followers. Leadership should be considered as a *process* where leaders and followers interact in the situation they are placed in, and it is this interaction that makes leadership more complicated but also more creative.

Leaders need to draw on a range of resources to convince and motivate followers to join in the journey toward a common goal. Engagement, initiative, and cooperation cannot simply be decreed, they are generated from the hearts and minds of followers throughout the organization. It's a game with three players, the leader, the follower, and the context.

Situational Leadership

The shift of focus from the leader to the leadership process led theorists to develop a 'situational approach' which considers the way leaders should adjust their style of leadership depending on the maturity and motivation of their followers and the prevailing situation in which they find themselves. The situational leadership model[7] recommends four leadership styles, *telling, selling, participating, delegating*, depending on the mindset of employees and the situation they are placed in. If you are in an operational environment with routine behaviors and low level of motivation, then you must use the *telling* mode and manage people with authority. At the other extreme, if you are dealing with autonomous researchers, you should move to *delegation*.

This approach has proved very popular and is extensively used in training and development, as it is simple to put into practice. But this model remains limited and does not consider, for example, the strategic competence of the leaders nor their ability to adjust the context and the culture of the organization to adapt to change.

[7] Hersey and Blanchard (1980).

Transformational Leadership

The *transformational* leadership approach corresponds to a much higher level of interaction between leaders and followers, far away from the standard transactional exchange that is generally observed.

Here leaders focus attention on motivating, guiding, and engaging followers to best achieve their personal goals and the aims of the organization. They transform followers, that is, help them evolve and engage, by adopting a strong role model, by sharing values and vision, by empowering everyone to achieve expertise in what they do and to innovate. This relational investment creates a bond that increases the motivation of followers and converts them into leaders. In a climate of trust and respect, followers transcend their own self-interest for the good of the group and they reach standards of performance which they would not have otherwise achieved.

Transformational leadership seems intuitively relevant because it takes into consideration the emotions, needs, and values of potential followers and it mobilizes their full selfhood. Transformational leadership is an ambitious approach that we have made more explicit with what we call the *creative model of leadership*, a collaborative model that shows how leaders mobilize the emotional support of their followers. But, before considering this model, let's examine charismatic leadership.

Charismatic Leadership

This is an interesting situation where the motivations of leaders and followers are fully aligned as leaders are able to resonate with the views already held by their followers and synthesize their aspirations. The followers promote the leader who embodies their goals and expectations.

Charismatic leaders are gifted individuals who can *identify* and *discern* key challenges and major stakes that exist in a latent state in the group and the specific context. They are carried forward like a surfer on the wave and become the representative of a deep movement which animates the group. They create hope by articulating a coherent vision that speaks to the collective imagination, because it fits the deep-seated expectations of the group. They are good story tellers, able to manipulate symbols, use metaphors to explain, and make the vison more explicit. They will pursue the vision to make things happen with a courageous and determined stance, setting the right example, walking their talk, and communicating with passion and inspiration. They are not afraid to take risks and persist in the face of setbacks.

In short, the charismatic leader becomes the amplified voice of their followers. This way of leading is the source of inspiring stories of exemplary leaders who make history, but it needs champions with innate qualities, listening abilities and a personal aura. Without violence, Gandhi led the destiny of a whole country, followed by huge crowds drawn by his capacity to embody their aspirations. Martin Luther King, carried forward by a great mass of followers, brought about the transformation of fundamental legislation in America. However, such inspired champions are not so common or easy to find and there remains the danger that charismatic leadership will evolve towards fusional leadership when leaders, intoxicated by their feeling of confidence, become less and less attentive to the expectations of the group and take advantage of their position of power to impose to obedient followers their own vision or a dangerous ideology. Mao Zedong, Adolf Hitler, and Bernard Madoff all led crowds of followers who identified with these deviant figureheads.

Leadership of Collaboration and Learning

The Interwoven Texture of Leadership and Followership

When James MacGregor Burns, the American historian and political scientist, defines leadership, he is keen to point out the importance of understanding the social aspect of power and the interwoven texture of leadership and followership.[8] Leadership is not merely a property of leaders but *emerges* from the relationship exercised between them and their followers. Leaders mobilize resources to arouse, engage, and satisfy the motivations and aspirations of followers. "In brief, leaders with motive and power bases, tap followers' motives in order to realize the purposes of both leaders and followers… Leadership, unlike naked power-wielding, is thus inseparable from followers' needs and goals." It is finding the way of engaging all actors to cooperate and learn together toward a common goal. The individual intelligence of all actors adds together into a collective intelligence to help the organization evolve, adapt, and survive in a highly volatile environment.

In summary, mobilizing large groups of people with different interests towards a common goal becomes a real challenge. Figure 1 shows an extended view of leadership, from the easy and comfortable way of imposing authority to the hard way of developing a creative process of collaboration.

Exploration and Execution

Leading and Managing Are Mutually Supportive

To adapt to a volatile and competitive environment, the creative leadership approach must combine two complementary modes of action: an *exploration mode* in search of valid strategic directions, and an *execution mode,* to make things happen. This dynamic, shown in Fig. 2, is essential to give the organization agility and adaptive flexibility.

[8] Burns (1978).

Exploring New Visions and Perspectives

Leaders are primarily responsible for defining and questioning the strategic direction and for taking their followers where they would not have gone on their own. They engage them on ventures they might have not otherwise considered. They make connections with the outside world and the various stakeholders to question the status quo and explore new perspectives. And this questioning leads them to accept taking some risks. In brief, they "show their people a future" as Napoleon used to say. This is a way to give followers some meaning to their action and a reason to engage in the chosen direction and contribute to its success. This contribution is not

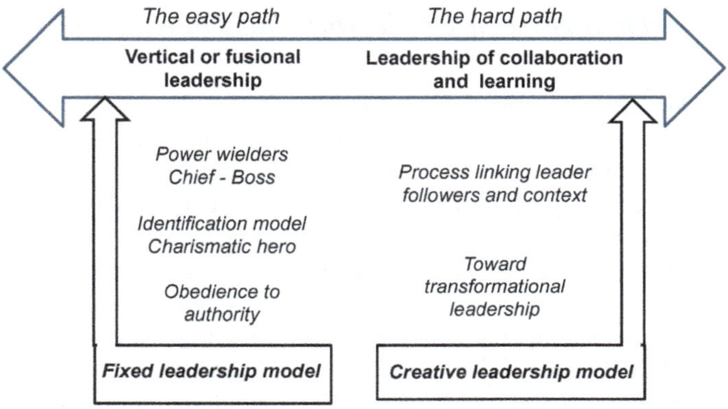

Fig. 1 Extended view of leadership

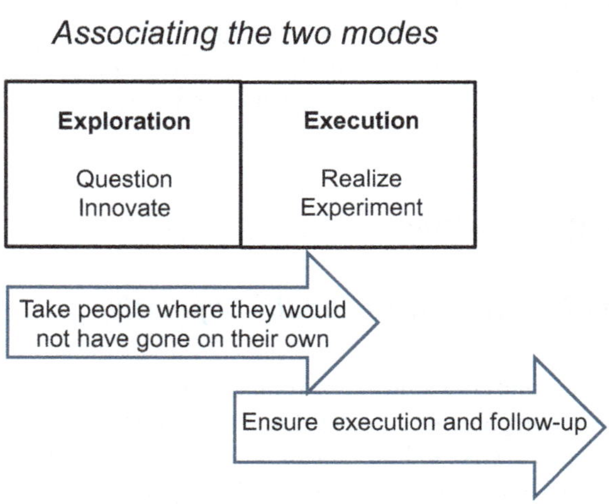

Fig. 2 The explore/execution dynamic

only economic but also an act of faith in the future, a feeling of doing something that will have an impact and can be a source of pride and self-esteem.

Execution and Experimentation

Being able to inspire followers, and telling them a credible story, is one thing, but acting and realizing the intended results is another. Even if leaders are successful in explaining and orchestrating their strategy, what matters and what remains to be done is to ensure follow-up and execution. They have to make things work and manage operations where the rubber meets the road. Leadership includes inspiration, as much as nuts and bolts. And if leaders don't focus on action, who will? As Herb Kelleher, the former head of Southwest Airlines, said: "We have a strategic plan, it's called doing things". All too often, leaders remain primarily concerned with getting their strategy down on paper without devoting enough time to implementation. An average strategy with flawless execution may be better than the reverse

This balance between exploration and execution can vary according to circumstances. In the life cycle of a product or a service, the emphasis is on exploration at the beginning of development, during the innovation phase and later moves to execution as the product or service becomes commoditized.

It is important to keep the balance between the *openness* of strategic questioning and the *closure* of execution, as it seems so much more attractive to become a heroic leader in search of new ventures than to be the manager responsible for day-to-day operations. Separating the seductive aspect of innovating from the banal aspect of managing operations can only lead to a dead end.

Would a team accept working for a manager who creative leadership of cooperation will remain flexible does not have the qualities of a leader? What about leaders who are not concerned with managing and implementing their strategy? Creative leadership of cooperation will remain flexible and adaptive by keeping the balance between innovation and action, by seizing opportunities when they arise at all levels of the organization, in all teams, in all meetings, and in open or closed networks.

Importance of Context and Culture

When the Context is not Right

All too often, after a leadership training seminar, participants return full of enthusiasm, ready to experiment, with a more open relational mind, only to come up against a context that has not changed. They are then back to short-term thinking, cumbersome hierarchies, power struggles, and transactional relationships. They cannot keep up their new behavioral style in a social context that is not supportive. Their naive and enthusiastic approach makes them easy prey for those who know how to play the age-old power game. The disillusionment leads them to give up or to become cynical where they could have contributed to a fruitful collaboration.

Creating the Proper Context and Culture

Actors placed in a specific situation react in their *own smart way* according to the organizational context (e.g., the hierarchical structure, the level of empowerment, and the appraisal systems) and the social culture (the shared values and rules of the game). Leaders therefore will have to adapt the systems and the cultural context to effectively mobilize followers. This can be done by flattening the vertical structure, by giving employees more autonomy and resources, and by freeing up their initiative closer to the operational level. Similarly, leaders can transform the socio-cultural context by introducing more fluid and trustful relationships, by changing the established rules of interaction and the incentives (rewards, constraints, or sanctions). For example, collaboration between players in a soccer team can be strengthened if they are valued not only for goals scored, but also for successful passes.

The Creative Leadership Model

From the Easy to the Hard Way

To sum up, great leaders should engage in a creative collaboration with their followers to unlock their potential contribution rather than control their behavior. They leave the *comfort* of the imposition of vertical power to take a less comfortable but adaptative approach. Management scholars and trainers have taken up the challenge, and there are thousands of publications, books, and seminars that have been devoted to developing theories and leadership models that promise, if not harmonious, then at least satisfactory cooperation.

An Explicit Leadership Model as Reference

Most prescriptive leadership models found in the literature can be described along three dimensions, the strategic vision, the relational style of the leader and the contextual culture that will motivate people and create an environment where they enjoy working and collaborating. We have integrated these three dimensions into an explicit model, '*the creative model*', a three-step process, that will serve as a reference.

- *Strategy.* A strategic long-term vision, open to questioning and exploration. In this way leaders assert their competence and engage followers by giving meaning to their actions.
- *Leadership style.* An engaged style of leadership to mobilize followers by sharing the vision and by empowering them to further explore and implement it.
- *Context & culture.* A determination to recruit the right people and place them in a context and a culture that will facilitate trust, collaboration, and learning.

This creative model appears quite convincing because it works with key motivations of followers such as autonomy, achievement, or social integration and also combines exploration and execution. Leaders and followers in close association realize a step-by-step transformation of the organization according to a continuous learning mode, each problem becoming an opportunity to question, improve, or innovate.

Nevertheless, this ideal model remains a *prescriptive mirage* which is difficult to implement. Many studies show that a very low percentage of employees are actively engaged in their job and enthusiastic about their workplace.

The Limit of Prescriptive Models

Leaders Still Disappoint

Despite the multiplicity of elaborate theories and convincing models of leadership, leaders regularly fail to satisfy their organizations, their employees, society in general and themselves. Most approaches remain prescriptive, as they merely set principles and give guidelines or advice. They remain at a high level of abstraction and too simplistic to produce lasting change as they skim over the complex reality of operations. To be effective they should operate at the level of the relevant actors placed in their prevailing context and culture.

In fact, leaders are easily swept along by the urgency of decisions, the speed of change or the strength of the competition. They easily revert to their natural inclinations toward centralization, vertical imposition of power, focus on immediate results or transactional culture. To get out of this regressive mode and engage their followers into a genuine culture of collaboration, leaders must work with application against their natural limitations and biases that operate under the level of consciousness. They also need to better relate to and motivate their followers. To achieve this, they have to learn about the way their brain functions, and this is where neuroscience can help.

Neuroscience can Help

We generally believe that we are rational creatures that behave consciously in a reasonable way. This is not the way our brain works most of the time. We are regularly subject to our affects, habits, and intuitive inferences, and our perception of the world is distorted by fear, immediate rewards, basic needs, and many other unconscious biases. To succeed over the long term, leaders must be aware of the predispositions that hinder their efforts to mobilize the motivations and aspirations of their followers. They have to work hard to overcome their limitations by learning about how their brains function, in the same way that outstanding athletes in the world of sports work hard to understand the functioning of their bodies.

Why Did the Taylorian Bureaucratic Model Spread so Easily

While prescriptive models often remain a mirage, the bureaucratic Taylorian model spread remarkably well in the wake of the industrial revolution and has continued to spread with some adjustments. When we interrogate its success, it appears that it was based on a rational and methodical approach to organizing operations, and on the imposition of a command-and-control pyramidal structure that ignored the basic motivations and aspirations of subordinates, who were simply considered as pawns. The Taylorian model took the *easy path,* and its effectiveness was gained at the expense of autonomy, cooperation, and personal development of employees. This rational approach did not take into consideration the basic motivators that are essential to engage followers in a fruitful collaboration. It thus remained strongly influenced by our inherent predispositions and biases which were allowed to operate freely.

So, let's see how organizations have evolved since the industrial revolution and the adoption of the Taylorian bureaucratic model, before considering what it takes to move to a more convincing leadership model.

References

Burns, J. M. (1978). *Leadership.*
Carlyle, T., (1841). *On heroes, hero-workship and the heroic in history.* James Fraser, London.
Diamond, J. (2019). *Upheaval: Turning points for nations in crisis.* Little, Brown and Company.
Hersey, P., & Blanchard, K. (1980). *Situational leadership model.*
Hofstede, G. (2010). Power distance. In: *Cultures and organizations.*
Milgram, S. (1974). *Obedience to authority.*

Chapter 2
A Straightforward Rational Managerial Model

Industrial Revolution and Mass Production

Division of Labor and Standardization

The Industrial Revolution resulted from a complete reversal of the craftmanship approach traditionally adopted to make a product. Craftsmen designed and produced bespoke objects based on the specific demands and desires of their customers, and they would maintain and repair these objects throughout their lifetime. For example, they would handmake a knife from beginning to end based on the demand of a specific customer. The Industrial Revolution transformed this approach by playing on two key concepts: *division of labor and standardization of components*. The knife was designed by a technician who decided on a shape that was imposed on customers. The knife was then broken down into components such as the blade or the parts of the handle, and each one was mass-produced. As the different parts were standardized, they could be produced in workshops far apart geographically. This would later result in globalization and the spread of suppliers all around the world.

As mass production gradually replaced human labor with increasingly sophisticated and fast machines, costs could be reduced with the increase in volume and automation. With the capital-enabled investment in machines and factories, capitalism and productivism were born. Mass consumption and consumerism emerged and grew. Price being one of the only ways to differentiate standardized and commoditized products, pressure on cost reduction intensified and led the famous economies of scale brought by the speed of machines, the level of automation, and the extent of globalization. Firms had to become the industry leader with lower costs than the competition and higher volume. Therefore, to avoid bankruptcy, manufacturers needed to find new customers drawn to cheaper and increasingly standardized products.

© The Author(s), under exclusive license to Springer Nature Singapore Pte Ltd. 2023

J. Teboul and P. Damier, *Neuroleadership*,
https://doi.org/10.1007/978-981-99-5122-2_2

The Scientific Organization of Work. A Rational Approach

As firms were left with the limiting option of increasing productivity and size to be able to offer the lowest price, they relied on the development of scientific methods to organize work. This rational approach, which separated conception from execution, those who think from those who do, was perfected by Frederick Taylor and his followers. They analyzed production operations in detail to determine optimal time and motion. Production planning and task analysis were entrusted to managers and experts, and operations were passed on to blue-collar workers. Technicians, engineers, salespeople, or accountants, who belonged to a new class of professionals, known as the white-collars, took autonomy and responsibility away from the blue-collar workers.

The Taylorian Approach Was Grafted onto the Bureaucratic Model

The Industrial Revolution initially gave rise to a primitive form of capitalism which nearly crumbled when faced with the anarchy of markets and unstable employment. Not only were workers not protected, but badly structured companies stood little chance of surviving. This capitalism, which threatened to lead to revolution, managed to avoid it by taking on the form of social capitalism. It established itself on a stable bureaucratic model, built to last and to survive the battering of the market. This method of organization placated the workers as they were now integrated into the workplace and prevented them from taking to the streets to protest their lot.

The Taylorian organization successfully fitted into the bureaucratic model based on the pyramidal structure. Like the traditional army, the pyramidal structure, that established hierarchical positions and well-defined functions, was able to organize huge crowds. As the effective control of a group could only extend to a limited number of individuals, let's say seven for example, effective power was shaped according to a rake structure of seven people supervised by a leader at each level. This produces large pyramidal chain of command. With four hierarchical levels, it is possible to manage 2500 people and with six levels 120,000 people.

Let's now analyze the Taylorian bureaucratic approach around the three dimensions that we have used to define the creative model of leadership: the strategic vision, the leadership style, and the relational context.

The Strategic Vision

In the Taylorian model, vision, plans, and budgets are conceived at the top and then broken down into multiple objectives that are deployed and *pushed down* through the

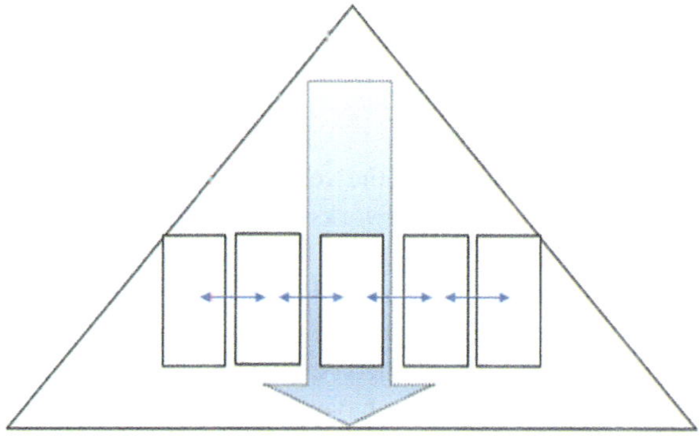

Fig. 1 Silos and transactional relationships between business units

hierarchical structure; the whole organization is thus able to focus on *fixed objectives and results*. Everything is done to achieve immediate productivity gains and tightly control operations, while ignoring longer-term adjustment to the customer's needs. In a similar way, the Gosplan, the state plan, in the planned economy of the USSR, decided which goods and quantities would be produced in all factories and businesses all over the country.

At each level of the pyramidal structure, managers focus on their local goals and are mainly concerned with the rational and technical aspect of their operations without seriously considering the motivations and the social interests of their subordinates nor the collateral effect of their decisions on the rest of the organization. As each unit gives priority to its own objectives, without worrying too much about the next or neighboring unit, activity and responsibility move up and down, within vertical silos, rather than laterally across units (Fig. 1). This formation of silos stifles the free flow of information transversely to deliver the final product or service. As each silo protects itself from the others, conflicts invariably appear between them in response to the variability of demand, in an increasingly volatile environment.

A Command-And-Control Leadership Style

Managers define the objectives to be reached and sell them to their subordinates when negotiation is possible, although usually they simply tell people what to do. This dry command-and-control communication system does not encourage collaboration and engagement, but it is well adapted to keeping the whole system under control. The *hierarchical bias* that we have already mentioned creates an internal competition to gain power and status and to rise above others, all the more as one gets closer to the top.

At the bottom of the pyramid, employees must conform and comply with the routines and procedures in place. As they have little autonomy and room for maneuver, they refuse to take initiatives. In case of difficulty, the first solution does the trick, or the problem moves upwards. Above all, they seek to do their work according to prescribed procedures and rules that they use at times to protect themselves against hierarchical pressure.

Interestingly, as power moves down the vertical pyramid, middle managers accumulate knowledge on how the system works, when to make exceptions, or how to interpret orders in order to oil the cogs of the bureaucratic machine. These managers find more meaning and gain more power by playing this useful role, and they can do it because the stability and the longevity of the system allows this type of learning. The result is that some distance gets established between power at the top and the effective control of action lower down. You may have formal power but little authority.

A Rigid Context

In the hierarchical pyramid each office, each element, has a defined function. Thus, the structure can be maintained regardless of who occupies each of the various cogs in the machine. Working in this kind of institution is like living in an "iron cage" as Max Weber puts it,[1] a rigid structure in need and search of stability. As long as the environment remains stable, the system functions well and can almost be managed on autopilot. In the event of a deviation from the norm, supervisors or controllers apply rational thinking to find the source of the deviation and correct it.

But, in a volatile environment, anomalies and incidents can arise at any time and it becomes essential to deal with uncertainty. This is why employees trapped by the rigidity of the system manage to recover some autonomy by holding buffer stocks, by refusing to push productivity and efficiency to their limits, or by retaining some hidden freedom of action. All these margins for maneuver that slip between operations lead to huge hidden costs, but no matter, customers will pay for them—at least as long as they have no other choice. Efficiency is eventually sacrificed to safeguard the stability of the organization and its long-term functioning. And when change happens, reorganizations are hated as they leave employees wondering where they will land in the new structure or whether they will still have a job.

A Transactional Culture

So, in this well-defined and predictable context, actors are not encouraged to interact and collaborate, and a *transactional culture* takes hold within the organization. The various players defend their own local interests and relate to others in a contractual and formal way with a low level of trust. People adhere to prescribed rules and codes, while ignoring a more interactive *relational engagement*. They clearly defend their

[1] *The Protestant Ethic and the Spirit of Capitalism,* 1904.

self-interest, as they are driven by their *egocentric bias*, an egoistic tendency that puts us at the center of our attention and makes us overestimate what we think or do. In the Taylorian organization everyone is driven to behave in a way that serves their self-interest. In the words of Adam Smith[2] "It is not from the benevolence of the butcher, the brewer, or the baker, that we can expect our dinner, but from their regard to their own interest."

Hiring

The recruitment of executives and white-collar staff is carried out on a technical basis, the rational aspect being privileged at the expense of relational skills. Blue collar workers, however, who have fragmented and repetitive tasks remain poorly qualified and easily replaced. Ray Kroc, who organized the McDonald's fast-food chain according to the principles of Taylorism, swore by the rule that any operator should be replaceable within two hours. He was especially keen to not be dependent on the oversized ego of the cook.

Learning is limited, and there is no resulting skills development when the environment evolves at high speed faced with the rise of new technologies and globalization.

Rewarding

So, what motivates employees to willingly enter this rigid framework, with such a loss of autonomy and meaning?

There is not much other choice than to play on extrinsic motivations such as salary and bonuses based on performance, with a higher level for the people closer to the top (outrageous CEO salaries). This choice corresponds to the fact that monetary compensation exercises a rapid and powerful stimulation of the reward circuit of the brain through the influence of dopamine. This neurotransmitter actively pushes us to grab all we can get, and even more (more on this in Chap. 4). This is perhaps one of the biggest design flaws of our brain, because the system of immediate reward is more powerful than the reasonable brakes that the rational brain can muster against it.

We call this predisposition that attracts us to an immediate gratification under the radar of our full awareness, *the bias of immediate reward*.

---

[2] Smith (1776).

Bias of Immediate Reward

Our brains have been shaped over millions of years of vertebrate evolution, by the necessity to survive and adapt to a chaotic and dangerous world. This led to the emergence of essential neural networks, such as the reward circuit, grafted onto the primary instinct of hunger, to stimulate the search for food in environments where resources were scarce and difficult to obtain. In the same way, the fear circuit and the corresponding bias of insecurity served us well. These networks are still part of the brain of today's Homo sapiens, and this hard wiring may lead to distorted and inconsistent behaviors such as, for example, obesity in a world where food is abundant. Hence, immediate rewards have a powerful and immediate effect but, as a drug, one needs more and more of it to keep the same level of stimulation.

Another extrinsic motivation results from the protection aspect delivered by the security and reassurance provided over time by the stability of the bureaucratic organization. It's a way to respond to the demand from the *bias of insecurity* that we have already mentioned. A third bias could also be mentioned, the hierarchical bias (described above), as some people are motivated to climb the ranks of the hierarchical pyramid and get power and status. This deferred gratification allows them to accept limited short-term rewards and prevents them from quitting.

But above all, the Taylorian model neglects the intrinsic aspirations like finding enough meaning and interest in the job. Followers are simply considered as pawns that can be moved at will.

Resilience of the Taylorian Bureaucratic Model

Despite all its shortcomings, the Taylorian bureaucratic model was able to successfully spread across the globe because its effectiveness was the result of a straightforward command-and-control approach based on rational methods, easy to represent and share. Leaders could operate without excessive effort as they relied on the hierarchical bias to exercise power and extrinsic motivators like carrots and sticks. Four cardinal predispositions could then operate freely:

The bias of fear and insecurity: As they feared being judged and disapproved by those in power or by co-workers, employees would avoid speaking up when they should, and refrain from questioning the choices and decisions from above. In a climate of psychological insecurity, they would not risk taking initiatives that could go wrong or be misperceived. To avoid being at fault, most would deny their mistakes or hide them. In fact, displeasing one's boss would carry much more weight than seeking to achieve good results.

The bias of immediate reward: Motivators like money, bonuses, and status have an immediate effect and remain extrinsic as people are swayed by them. They are not intrinsic as the impulse does not come from inside the person. Moreover, long-term rewards are easily devalued.

The egocentric bias and self-interest: Egocentrism corresponds to the mental image of oneself, of everything related or associated with oneself. People are preoccupied with their own interests first and foremost, before trying to collaborate with others. This leads to a climate of distrust.

The hierarchical bias: We have seen that the exercise of vertical power operates according to this bias that is deeply wired in the neural circuits in our brains.

The effectiveness of the Taylorian model has produced stress and demotivation at work and has proved painful in terms of individual freedom and development, but many organizations still operate on that model despite its multiple disadvantages and the lack of motivation it entails.

The Fixed Model of Leadership

From the Taylorian model it is possible to imagine what we call a *fixed model of leadership* where people governed by their natural predispositions passively accept authority and are not motivated to go beyond planned objectives and defined roles. This fixed model will serve as a reference to be compared with the creative model along the same three dimensions. The model is based on the most regressive situation that can be observed in each of the three dimensions. A strategic vision channeled by the imperative to obtain immediate and tangible results, a leadership style that shows a lack of commitment and empowerment, and a cultural context where complexity and transactional interactions rule without trust and cooperation. So, the result is a limited, restrictive, and rigid model with a Taylorian overtone.

Three Main Dimensions of the Fixed Model

The strategic vision, designed by a few top managers and executives that have access to comprehensive information, is channeled by the imperative to obtain immediate and tangible results. The planned and fixed objectives, often expressed in financial terms, remain abstract and fade away by the time they reach the operational level where they lack meaning. They do not permit the necessary questioning and adaptation in a changing environment. This default model is not able to generate a global and collective meaning to individual action. The overall approach is disjointed and lacks coherence, especially after necessary restructuring. Employees are stressed and feel confused by disruptions and unpredictable reorganizations.

The leadership style shows a lack of commitment and empowerment with the return to centralization. Delegation only takes place with multiple checks. The struggle to gain status and power intensifies in a competitive climate. The vertical exercise of power and the internal culture of competition undermine the emergence of a climate of trust and cooperation.

The lack of an enabling context and cultural framework. Hiring remains local and is not a key strategic decision. Motivation remains extrinsic, and the collective effort is not rewarded. Rewards remain individual. The context is difficult to read in a complex multilayered structure that does not favor collaboration and horizontal coordination. The transactional culture limits relational engagement and learning and does not encourage informal trust.

This model is extreme in the sense that it is based on the least effort across the three dimensions with vertical exercise of power, lack of empowerment, extrinsic motivation, and people governed by their natural predispositions and biases. By positioning the fixed model and the creative model at the two ends of a spectrum, it is possible to display a range of intermediary situations between these two contrasted models (see Fig. 1 in Chap. 1).

A Range of Possibilities

To avoid drifting back toward the fixed model and to foster a more convivial and transformational approach where it feels good to work in collaboration, leaders will have to work with determination and discipline along the three dimensions, with varying degrees of success. Some will manage to find a balanced successful strategy for the long term, while still not abandoning the transactional culture, while others will engage in a wealth of opportunistic disjointed projects that will succeed thanks to a strong culture of relational engagement—but it is rare to see a coherent effort across all three dimensions.

So, if we want to move away from the reductive vision of the fixed model, we have to understand what restrains leaders from implementing an open and creative leadership approach. It is about time to understand how our brain operates and get a better idea of its capabilities and limitations.

Reference

Smith, A. (1776). *An inquiry into the nature and causes of the wealth of nations.*

Chapter 3
A Brain in Three Dimensions

Why Do we and Other Animals Have a Brain?

Clearly, not every species of living thing on our planet has a brain—trees and plants, for example, do not—so what is our brain for? To say that it enables us to perceive the world and to think may not be acknowledging the most important reason.

Having a brain enables us and other animals to move, to perform adaptive and complex movements crucial to our survival and reproduction. Trees and plants do not have brains because they do not move.

To be able to walk and avoid obstacles in our way, our brain constructs stable virtual representations of our environment by using information from our senses. These internal maps determine how we see the world and tell us how to move around. The properties of the external world are embedded in our neural circuitry.

So, our brains create our perceived reality, and we do not see the world as it is but as our brains represent it through virtual models. The brain then uses these models to anticipate our next move and help us move around. The brain is, essentially, an amazing *prediction machine.*

Vertical Integration: The Triune Brain

We start our exploration of the brain with the highly simplified model of three metaphorical layers that illustrates how the brain has evolved in complexity over time (Fig. 1). The three parts of the triune brain—the instinctive brain, the emotional brain, and the cognitive brain—are layered vertically on one another and refer to the evolution of the human brain in three stages. This long-held representation is no longer accepted today because these different levels have continually been reorganized and reshaped over the course of evolution. In humans, they work together in an intricate way. However, this approach is still a good way to start our exploration of this highly complex organ. It raises awareness about the long evolutionary history

© The Author(s), under exclusive license to Springer Nature Singapore Pte Ltd. 2023
J. Teboul and P. Damier, *Neuroleadership*,
https://doi.org/10.1007/978-981-99-5122-2_3

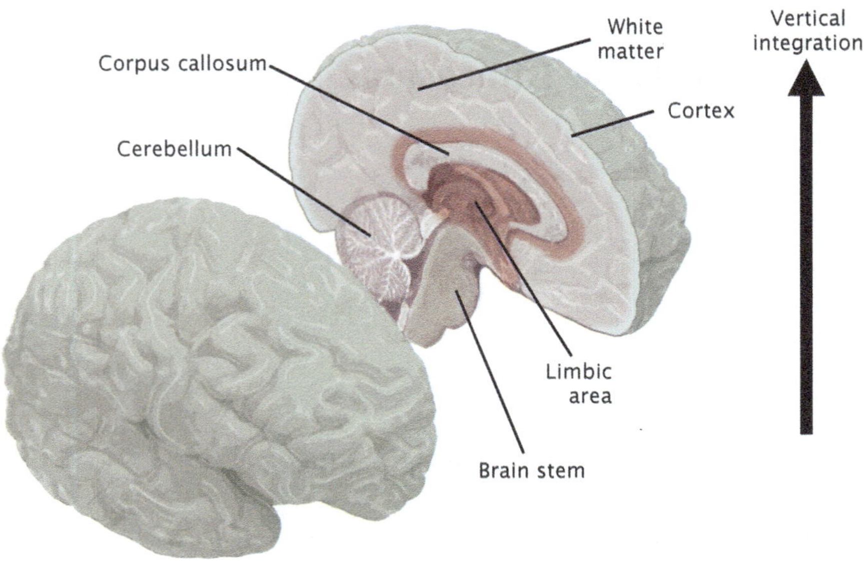

Fig. 1 Vertical integration. *If flattened out, the cerebral cortex would comprise a thin sheet of gray tissue of about 6 mm (6 layers of cells) that covers an area of about 800 cm*^{2}*, the size of a table napkin. To occupy less space, it is massively wrapped and folded so it can fit into the skull. Between the cortex and the corpus callosum, the white matter represents the massive interconnections of neurons*

of the human brain which still exhibits some functioning close to that observed in other animal species.

The oldest part—the brain stem and cerebellum—constitute the instinctual brain, also called the *reptilian brain* because it regulates automatic reactions throughout the body and contributes to its survival. Layered over it is the emotional brain, also called the *limbic system* which mainly deals with emotions and integrates rewards to facilitate learning. The third layer is the *cognitive brain,* a massive cerebral cortex in charge of perception, abstract thought, language, cooperation, and restraint of behavior. It is associated with the development of consciousness.

The Instinctual or Reptilian Brain

The spinal cord that brings in data from all over the body enters the skull in an area called the *brain stem.* This is the essential part of the brain of reptiles and fish, which depends on the contextual environment. For example, the temperature of their body is determined by the external temperature. The brain stem helps regulate our basic physiology, temperature, breathing, and swallowing. It also directly controls our states of arousal and our drives such as hunger or sexual desire.

For humans, the brain stem, in relation with the limbic area and the cortex, oversees safety. We react instantly to a situation of danger, with the fight-or-flight mode of alertness. The body readies for action, adrenaline pours into the blood stream, cortisol is released, and our metabolism is prepared for the energy demand that will be required. Then, if we feel helpless, we freeze or pass out, simulating death.

The Emotional or Limbic Brain

The limbic region of the brain markedly developed with the arrival of mammals 200 million years ago to allow emotional attachment and a fuller expression of emotions. It turns sensory perception into emotional and physical responses and assigns value to whatever situation or object is encountered. 200 million years of evolution have made the limbic system a refined instrument capable of making fast decisions with little information and of learning complex behaviors. Two key elements of this region of the brain are the amygdala that is part of the fear system and that reacts to emotionally relevant stimuli and threats, and the ventral striatum that is involved in motivation and is crucial to learn adapted behaviors from experience.

The Cognitive or Cerebral Brain

The third level is the cognitive brain. It corresponds to the development of the outermost area of the brain known as the neocortex or new cortex. Appearing when we became mammals, it gradually expanded and spread to surround and cover the old limbic system from the back of the head to the front. Our extraordinary capacities to think, love, and understand our universe depend mainly on a thin, greyish veneer of about 6 mm thick under which we find bundles of long-distance connections that form the white matter.

As a result of evolution, this area increased in complexity and became organized into a huge number of modules, each with a different goal. The part that has increased most dramatically in humans and that makes us different from other primates is the frontal lobe, the rational part of the mind. It deals with abstract thought, language, decision making, and empathy.

Rational thinking, which we are so proud of, has its roots in the archaic brain where powerful binary primitive impulses operate such as approach and avoid, good or bad, friend or foe. It is only in the upper layers of the brain that different nuances of grey appear.

Cognitive Externalization

For tens of thousands of years humans only had their brains to preserve information. While the brain has a large memory capacity with its multiple long-term memories, the emergence of written signs and numbers opened the immense field of external recording, mathematical computing, and mass processing of data. 5000 years ago, some brilliant Sumerians came up with the idea of a system of notation that enabled facts to be recorded, censuses and numerical calculations to be carried out. This was the beginning of the administrative organization of the economy: it became possible to record information in written archives using armies of scribes, then office workers, librarians, and accountants, and now, battalions of servers in the cloud. It should be noted that access to these records on shelves or in computer memories is different from the functioning of the brain's working memory, which instead operates by jumping from free association to free association, in a tree structure. Advances in miniaturization have increased this cognitive externalization. We now have instant access to huge registers of information with connected devices that fit in our pockets.

Similarly, the Taylorian model came to rely more and more on cognitive externalization to organize operations. The learned skills, mainly held in the craftsman's brain, were externalized into models, mathematical formulations, norms, and procedures. This process paved the way for artificial intelligence.

Left–Right Integration

The brain is made up of two cerebral hemispheres, the left brain and the right brain, connected by the corpus callosum, a large bundle of fibers that transmits information between the two hemispheres.

Over human evolution, the left brain and right brain have developed separate but complementary functions. With the growing demand for cortical space and the development of brain structures that facilitated language on the left side, brain capacity expanded further by reducing redundancy.

In the 1970s, popular culture viewed the left hemisphere as the logical brain, devoid of emotions and rather boring, and the right hemisphere as the creative brain and the "cool" side. This oversimplification has clearly been overused and has often served as the basis for defining personality types or artistic talent. Recent research has shown that brain asymmetry is more complex, but the distinction between the hemispheres remains valid and useful. To illustrate this distinction, we contrast the working of the two sides of the brain on some key areas (Fig. 2).

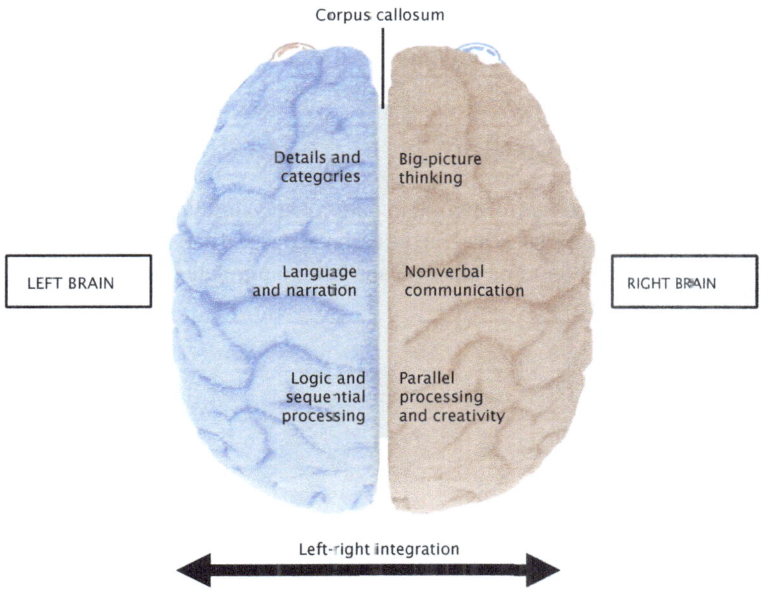

Fig. 2 Left and right brain

The Right Hemisphere

Global Properties of Forms and Objects

The right hemisphere is about big-picture thinking and holistic reasoning. It allows us to grasp the context of something, the global perspective, and to make sense of the whole. It is well adapted for dealing with images that can condense a huge amount of information. Thoughts are often bolstered and enriched with pictures that assemble all the elements of a situation and show some of their relationships.

Nonverbal Communication

The right hemisphere deals with the emotional functions of language and nonverbal communication, that is, the prosody, the intonation, and the facial expressions. By extension, the right side can note inconsistencies between the tone of voice and the actual message or between body language and facial expression. These inconsistencies can signal when a person is not telling the truth.

From the beginning of life, we communicate with one another nonverbally. When we were babies, nonverbal signals were our lifeline, the only way we could convey our needs and wants. We became attached to our caregiver(s) through these nonverbal patterns. Babies have a larger right hemisphere until the end of their second year because it is only then that the left hemisphere begins its growth spurt. In fact, the right hemisphere dominates the brain for the first 3 years of life.

Parallel Processing and Creativity

The right brain can process information in parallel and simultaneously. It is thus able to make unlikely connections, which is the essence of creativity. It helps us appreciate humor and jokes, understand metaphors, and bridge unrelated ideas.

The Left Hemisphere

Details and Categories

People using the left side of their brain thrive not just on details but also on details of details. They analyze and break things down into manageable elements. In so doing, they categorize information into hierarchies or likes and dislikes. They will, for example, describe and judge how things are categorically different (e.g., name the parts of a flower or dissect the image of a rainbow). They label the perceptions processed by the brain and organize them into lists. To simplify, we could say that when we see a forest, the left brain focuses on trees and leaves and the right brain on the whole forest.

Articulation and Comprehension of Language

For most people, language is processed in the left side of the brain. The Broca's area, which is close to the motor cortex, deals with articulation and spoken language; the Wernicke area, at the crossroads of word recognition and speech comprehension, is situated in the rear part of the brain. When we speak, we operate in sequences, assembling words and strings of words to create sentences. It is interesting to note that speech processing on the left side of the brain seems to be rather literal, without prosody or intonation, aspects that are dealt with by the right side.

Logical and Sequential Reasoning

Most education deals with left hemisphere skills such as reading, mathematics, geography, all sorts of sciences that involve sequential reasoning, and thinking along linear chains of logic with the use of symbols. The left brain favors linear, logical, and systematic thinking. We take notes in meetings and organize presentations in a sequential way: A1, A2; then B1, B2; and so on. We solve mathematical problems with logical chains of statements.

The Left-Brain Interpreter

In the left hemisphere of the brain, there seems to be an interpreter module that helps us find order in chaos and fit things into a coherent story. It is driven to infer cause and effect or to guess an underlying structure, even when no pattern exists. It grasps the first and easiest explanation by using the available information. The interpreter will catch the gist of a situation, try to explain it at all costs, and reject information that does not fit in. It seems to be the bossy, know-it-all side of the brain, which often overpowers the right side.

Back-to-Front Integration

From Perception to Abstraction

The posterior part of the cortex is dedicated to perception by our senses and is organized into regions that specialize in different tasks, such as vision, hearing, and touch. The front part of the cortex is devoted to execution, from movement to abstraction. These two regions are separated by a great vertical fold called the *central sulcus*.

Perception

Each sensory region is organized hierarchically. Raw sensory information reaches the cortex in a primary projection area and is then sent to a series of specialized structures where it is deconstructed and analyzed before being further integrated into a stable representation in higher regions of the brain. This eventually leads to a response from the frontal cortex with movement and action.

Frontal Cortex

Beyond the posterior cortex responsible for perception, the frontal cortex in charge of thinking, planning, and initiating action but also relaying reason and emotion.

In humans, the frontal cortex represents a third of the surface of the cortex compared with a tenth for our near cousin, the chimpanzee. It makes us uniquely human. Being connected to all parts of the brain, the frontal cortex makes us do the more difficult things, thinking and acting, while consuming a lot of energy. It enables us to project ourselves into the future by planning long term.

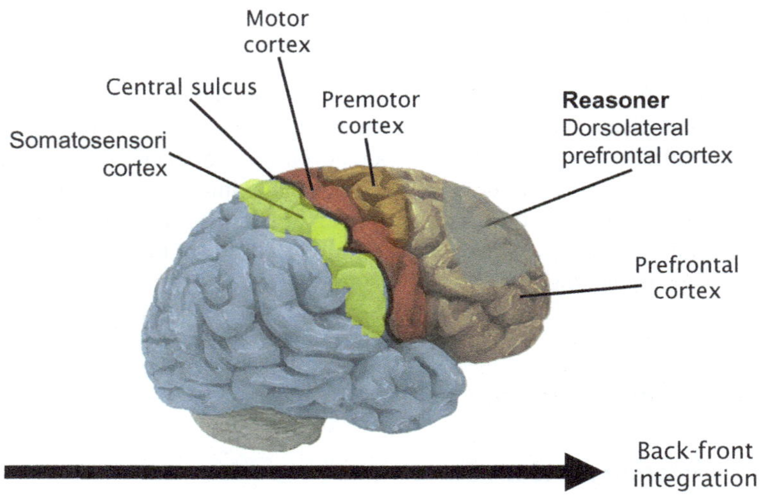

Fig. 3 From perception to movement and then to abstract thinking (Adapted from Baars & Gage, 2013). *Sensory and motor maps on both sides of the central sulcus (the central groove) represent our body in miniature. On the left side of the central sulcus are the somatosensory maps of the body and on the right side of the central sulcus, the motor maps. From left to right, the frontal cortex first deals with motor and premotor action, then thinking becomes more and more abstract toward the frontal pole*

Prefrontal Cortex

The prefrontal cortex is situated in front of the motor and premotor areas, as can be seen in Fig. 3. In connection with all the other regions of the brain, it makes us transcend automatic behaviors and choose which tune to play, focus attention on sensory inputs, plan or execute actions under control, produce abstract ideas, or regulate emotions.

The prefrontal cortex performs the most advanced and complex functions, the so-called ***executive functions***, which correspond to purposeful action after assembling bits of information and looking for patterns. It helps decide between conflicting options like speaking up or remaining silent, pulling the trigger or not. It can produce thoughts and abstract representations and lead to concerted action by fixing objectives and implementing them. It oversees self-discipline and rules: "don't smoke in the conference room, don't take this extra slice of cake, be sure to save for your retirement".

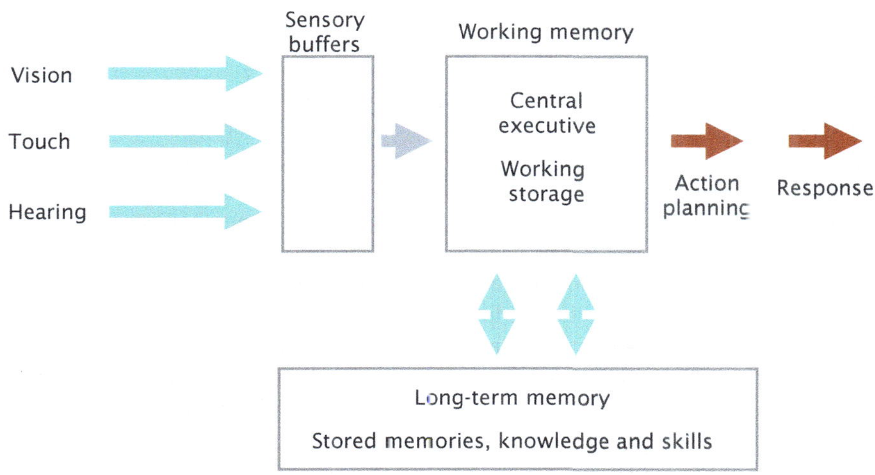

Fig. 4 Working and long-term memory (Adapted from Baars & Gage, 2013)

From Perception to Conscious Processing and Action

The activity of the brain can be represented as shown in Fig. 4. Broadly speaking, the back of the cortex oversees most of the work dealing with sensation and perception, as represented on the diagram with perception and sensory buffers on the left. Perception stimuli then access the frontal cortex to prepare action and thought in the working memory, mainly located in the prefrontal cortex. Working memory represents a set of mental processes temporarily accessible for use when striving toward current goals and engaging different parts of the brain.

The working memory, which prepares for action and underpins mental work, consists essentially of an executive center, the *central executive*, and a *working storage system* that keeps information online, to make it available to the rest of the brain.

Central Executive

Metaphorically, the central executive function is a kind of central administrator, that frees us from the automatic flow of experience and enables us to prioritize goals and plan action to adapt to our environment. It is crucial to our everyday activity, from carrying on a conversation to planning a holiday or driving a car, and it is close to what we mean by the "self" and our sense of free will. Because we are only aware of our conscious thoughts, we have a subjective sense of autonomy and control with a strong need to believe in it.

Working Storage

Working memory allows us to process conscious information in real time. It retains and updates multiple pieces of information in a short-term storage area with limited capacity, the content of which is constantly updated several times per second. Information disappears quickly if it is not considered important enough to be transferred to long-term memories. For example, we can keep a phone number long enough in mind to dial it on a keypad. When we read a sentence, we have to hold it in our minds, from beginning to end, to understand its meaning. Thus, everything goes through a narrow evanescent storage, but this limited storage capacity is compensated by the fact that we can easily retrieve information stocked in other memories of the brain. We can almost immediately recall the itinerary and the names of the cities we visited during a past trip or all the details of our wedding day. This remarkable efficiency has one serious drawback: the operation of the working memory is laborious and consumes a lot of energy. Moreover, given its limited capacity, it can only do one thing at a time.

The Magic Number Seven

This limited processing capacity seems to explain why the number seven is considered a magic number,[1] why phone numbers are usually limited to seven digits, why there are seven colors in the rainbow, seven days in the week, seven musical notes, seven dwarfs, seven wonders of the world, seven deadly sins, seven seas, seven lives, seven skies, seven arts, seven samurais, seven plagues of Egypt, seven arms of the Jewish candelabra, seven basic attributes of Allah, seven reincarnations among Buddhists, and so on. In practice, we usually only deal with three or four things out of the seven that are readily available.

The Theater Stage

It is quite insightful to use the analogy of a theater stage lit by the spotlights of attention to illustrate the conscious activity of the brain represented in Fig. 4. On the stage, the conscious processing of information in the working memory is represented by a few interacting actors (Fig. 5). This scene is where we visualize action and decisions, where we talk to ourselves, where we can freely imagine all kinds of

[1] A good illustration of this limit is the illusion of completeness, which means that when we read a page of text, only the few words of the sentence under our gaze are activated in working memory. The rest of the text is reconstructed from other brain memories. The same is true of Sperling's experiments, which show that an average of five letters are remembered from a table of twelve letters which has been quickly shown to subjects Naccache (2020).

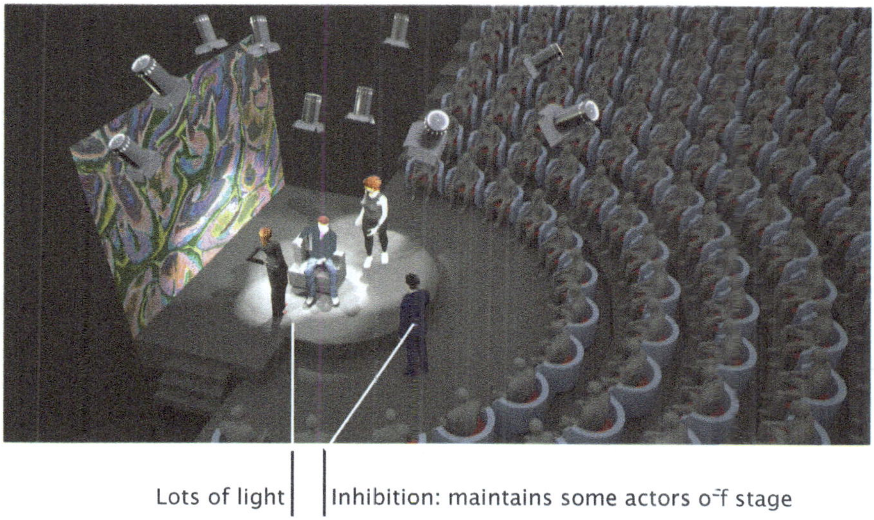

Fig. 5 The stage of consciousness

things like solving a math problem or calling our favorite restaurant. And the small part of the stage illuminated by the attention's spotlight illustrates the limited capacity of our working storage. The powerful projectors that illuminate the stage illustrate the working memory's great consumption of energy.

The stage of awareness is a powerful metaphor that illustrates a striking paradox: the limited processing capacity of our working memory, but its amazing ability to address any part of the brain. And it follows that our rationality remains limited even if it is enhanced by external systems and supports.

Thus, focusing attention on mental activities, such as performing a calculation or planning an action, puts other cognitive tasks on the back burner (imagine trying to perform a complicated calculation like 293 × 37, while simultaneously reciting the names of all the states in Africa). The brain must consciously focus on the task and proceed slowly, step by step.

As a result, when we consciously think, analyze a situation, or develop arguments to make a decision, we use our prefrontal cortex and proceed step by step while considering a limited number of options at a time.

The Rational Brain in the Dorsolateral Prefrontal Cortex

An important part of the prefrontal cortex is *the dorsolateral prefrontal cortex* which corresponds in great part to the rational brain. This region is involved in logical reasoning and cognitive control. Located toward the top and lateral sides of the frontal lobes, it resembles something like a hat over the prefrontal cortex.[2] It is the part that has evolved most recently over the course of our species' evolution and is the last to fully mature in adolescence, which can explain the difficulty teenagers sometimes have in controlling their behavior. We call it the "*reasoner*" because it is the center of logical reasoning and cold cognition.[3] By bringing together most of the information processed in other brain regions, it can choose to focus attention on a specific goal (such as describing a flower in detail, or focusing on reading a text), or to make conscious choices (such as cycling rather than driving) or plan for longer-term actions (such as planning a career change or inviting a friend over).

The reasoner can inhibit choices driven by immediate rewards to favor choices that will depend on future considerations. It is also associated with impulse control and social norm compliance (e.g., stopping us from making embarrassing confessions or stealing from a store).

We Have to Revisit Our Narrow Vision of Rationality

The ability to call on the reasoning center in our prefrontal cortex is unique to humans and plays an essential role in the scientific and technical world that we inhabit. Rational comes from the Latin "ratio", which means to measure and count. The result is that we have put rational thinking on the driver's seat and granted a central role to reason in directing our existence. But should the success of reason come at the expense of emotions and affects, and can reason be used to master our passions and irrational behaviors?

Current research in neuroscience shows that rationality is overstated and that we must put an end to this idealized vision of a human mind, capable of controlling its destiny by the force of reason. In fact, the brain works primarily with affects and emotions, and reason, which is recent in evolutionary terms, is built on the foundation of affects. We must not let science, technology, or artificial intelligence conceal the human part of this deep emotional functioning. The human being is, above all, a motivated being.

References

Baars, B., & Gage, N. (2013). *Fundamentals of cognitive neuroscience: A beginner's guide.* Academic Press.

Naccache, L. (2020). *Le cinéma intérieur.* Odile Jacob.

[2] Of course, complex neural circuits mainly activate the regions we mention.

[3] Cognition is the ability to orchestrate thought and action with internal objectives.

Chapter 4
A Very Emotional Brain

Revisiting the Notion of Rationality

Rational–Emotional Dichotomy

The triumphs of the industrial revolution and the world of technology celebrate our rationality, our capacity to plan the future and to act logically. The Cartesian approach, in accordance with the preeminence given to reason in Western thought, tends to categorize elements in opposition: reason and emotion, rational and irrational, logic and gut feeling, the clarifying light of reason and obscurantism. Reason, which makes us uniquely human and allows us to escape our "primitive" animal condition, is expected to control our passions and irrational behaviors.

Plato believed that the soul was divided into reason—which seeks truth—passions, and appetites (hunger or sexual drive). Guided by reason, we should, like a chariot driver, master two impetuous and ill-assorted horses, one preoccupied with recognition and glory, and the other with primary urges. Reason must maintain the balance between these two urges. Hence, Plato believed that the benevolent dictator should be guided by reason to lead citizens and not act by manipulating emotions. Gustave Lebon, in his studies of crowd psychology, saw mobs as gullible and capable of dangerous and emotional behaviors. Rationalist thinkers believed that humanity could gradually free itself from passions, superstitions, and all kinds of irrational thinking.

We are not expected to express emotions in social situations or at work and we are more at ease talking about them than feeling them. Many managers even think that emotions have nothing to do with decision. They do not allow themselves to listen to their emotional or social intelligence. They prefer to refer to solid and objective data such as balance sheets, returns, or financial forecasts that can be processed rationally. And yet emotions are never absent.

By favoring the abstract world of invented signs, we take the risk of cutting ourselves off from sensory inputs and the contribution of our emotional life. By privileging rational approaches, academic education in many professions does not give

© The Author(s), under exclusive license to Springer Nature Singapore Pte Ltd. 2023
J. Teboul and P. Damier, *Neuroleadership*,
https://doi.org/10.1007/978-981-99-5122-2_4

enough importance to soft skills, which are so essential for the proper management of human relations and leadership roles. As a result, we remain ill-prepared to realize how much we are conditioned by our unconscious emotional functioning and how affects and emotions play a key role in our lives.

According to David Hume, the great Scottish philosopher of the eighteenth century "Reason is and ought only to be the slave of the passions and can never pretend to any other office than to serve and obey them".[1]

Current research in neuroscience shows that the dichotomy between reason and emotion is illusory and false. Our brain is not designed to produce purely rational behaviors that are devoid of emotions. Thought is not detached from internal affects and emotions. Without them we could not think, we could not act, we would not know if we were right or wrong.

We are not, from the outset, rational creatures deciding how to act without worrying about what we feel. We must put reason back in its place and see how it associates with emotions. Our brain is programmed to stay tuned to the functioning of our body and its affects, to assess the value of certain choices. Whether we are hesitating between two dishes, two job offers, or two investments, we see the world through affect-colored glasses.

Not only do emotions not hamper reason, they are essential to its functioning. Affects, passions, and pleasures drive Plato's chariot, while rationality remains as a passenger. The brain in fact works primarily with affects and emotions, and reason, which is recent in evolutionary terms, is built on the foundation of affects. The human being is above all a motivated being.

The Emotional Brain Plays a Crucial Role in Reasoning

Antonio Damasio's studies of the relationship between emotions and reason marked a turning point in our understanding of the crucial role of emotional evaluations to guide reasoning. His research showed that cognitive and emotional processes typically work together and are closely connected. Decisions are influenced by emotions, which prepare us to evaluate and filter what happens around us.

Damasio examined patients who had suffered a deterioration of key elements of the emotional network of their brain without affecting their cognitive functions. They were intelligent and their rationality still functioned well, but they were no longer able to assess emotionally what was happening to them. They lacked access to the emotional evaluation of things and events. Although the building blocks of rationality were still in place, they showed little interest in their emotional and social behaviors and were unable to concentrate on what really mattered.

---

[1] Hume (1740).

Elliot

In *Descartes' Error*, Damasio described Elliot, a patient who had a tumor removed from his orbitofrontal cortex.[2] Elliot was intelligent, well informed, and well mannered. However, after surgery, he was unable to lead a normal life. He started doing irresponsible things and became easily distracted. Each decision would take ages because he could not make up his mind between options. For example, he might spend over 30 min trying to decide what date and time would be optimal for his next appointment. Without an emotional evaluation of options, he continued to hesitate between alternatives. His rational mind looked fine because he could solve problems and recall events, but he was incapable of showing emotions. In fact, he could not live a normal life using only his rational brain. Guided only by reason, he behaved unreasonably.

As we do not have access to the internal functioning of our brain, we do not realize how much our mental reasoning depends on neural processes that operate below the level of consciousness. Even when our reasoning seems cold, objective, and logical, it is imbued with emotions, beliefs, or interpretations. Our brain is not naturally inclined to produce purely rational behaviors. Rationality is an additional construct, the icing on the cake of affects and emotions. Cognition (the ability to reason and make decisions) and emotion are not separable.

A Useful Brain Valuation System

Our sophisticated emotional system, which has been refined over millions of years, plays a major role in guiding our behavior with the generation of predictions and anticipatory patterns to help us move around and prepare for the future. But for prediction to happen, we need to appraise our internal states and our environment with a *value system*. We feel the world before we think. Every creature, even a one-celled organism without a brain, must possess such a valuation system to orient itself. Value signals tell us what is agreeable or unpleasant with a level of activation from calmness to arousal. It is the pleasure of feeling the warmth of the sun on one's skin and the intensity of the sensation.

These affects which assign value to things are not fully perceived at the conscious level. They initially play on binary oppositions such as good/bad, correct/false, gain/loss or strong/light, which could lead to an immediate response. Then they combine with more complex sensations to reach our conscious register and make more nuanced assessments to signal our intentions and prepare appropriate action.

---

[2] The orbitofrontal cortex is situated in the lower third of the prefrontal cortex, at the bottom of the frontal lobe and just above the orbit of the eyes (hence its name).

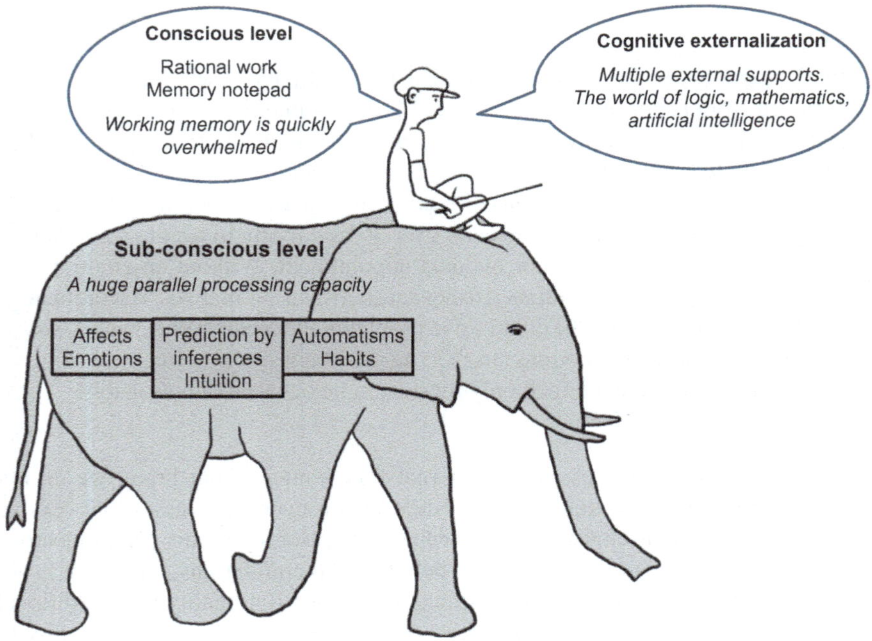

Fig. 1 Conscious and sub-conscious functioning. *Affects and emotions as well as habits*[3] *and intuition*[4] *are below conscious level*

Affects are the result of a continuous internal process called *interoception.* At any given moment, our inner somatic states based on pleasure or displeasure, arousal or calmness are detected and analyzed in some regions of the *insular sensory cortex* (see annex 6). All sensations of our internal organs and tissues, from the pit in our stomach to the accelerating beat of our heart, are represented in the insular cortex where the interoceptive activity is simply a summary of the state of our body in search of its balance. When they reach the conscious register, these feelings are then represented in the form of emotions. Most often, they remain as background noise that is still able to influence us.

In Fig. 1, the subconscious work of the brain has been represented as an elephant and the reason with a rider perched on its back, guiding the animal in a specific direction. But the rider depends on the goodwill of the elephant, that is, on the underlying results provided by the valuation systems beneath.

Without these internal affects, we would be like a sailboat with a rudder, but with no wind to move us forward. The elephant may have an emotional urge of its own, and we may have to work hard against this urge to reach our goal. If we can't succeed, we are quick to invent rational reasons to explain the unexpected move or find ways

[3] For habits refer to annex 7.

[4] We deal with intuition in Chap. 7.

to correct it at a later stage. We know how to use our reason to get an illusion of control and justify our actions post-hoc.

In fact, we don't have direct control over our affects with our rational reasoning because interoception and affects are at the very basis of all our perceptions and thoughts. But we can take a step back to observe and integrate them when they reach consciousness. Reason seems to be above emotion because it can be observed and elaborated, but motivation affects are irrevocably woven into the fabric of decision making. Emotion is *the energy* that drives, organizes, amplifies, or dampens cognitive activity.

What is the Point of Emotion?

Emotions are Action Programs

Emotions are quick, spontaneous, involuntary action programs carried out in our bodies to signal our intentions and enable us to rapidly adjust. They appeared during evolution where it was necessary to find food or face multiple dangers in a chaotic environment. They are triggered by 'competent' stimuli, based on the level of potential threat or satisfaction. They communicate important messages through a variety of facial expressions, body language or internal changes (cardiovascular or hormonal).

Each emotion has evolved to handle a specific problem: the imperative to avoid danger and pain, the search for rewards such as food, the importance of being included in a group, or the pleasure of being anchored in the here and now and enjoying the moment. Emotions color objects, places, and events with fear or pleasure, admiration, or disgust, and they command bodily reactions such as smiling or avoidance, approach, or revulsion.

Emotions are therefore a powerful source of motivation, which direct our behaviors towards whatever maximizes our own interests or the interests of our loved ones (the words emotion and motivation have the same Latin root, 'movere' which means to move). As such, emotions provide a compass that guides us toward or away from things or people according to the positive or negative associated value.

Emotional Intelligence

Intellectual prowess is not enough. Good leaders should draw on their emotional intelligence by processing emotional information skillfully in association with a rational approach. Affects and emotions are always there whether they like it or not. This means that to make good decisions, rational and logical methods must be integrated and combined with the leader's emotional experience in an active dialogue.

Emotional intelligence simply results from the fact that we must now recognize that we have an emotional brain and that we are able to understand, manipulate and regulate our inner emotional states with self-awareness and self-management.

Self-awareness is necessary to assess our personal drives, limits, and biases and to understand our inner beliefs and values. Being aware of how we feel and function emotionally enables us to validate and prioritize our choices. The objective is to become more assertive and self-confident.

Self-management is about self-control, managing our emotions, and staying calm under pressure. It also involves our ability to align our actions with our values and beliefs. The objective is to become more adaptive and ready for surprises and change.

The main difficulty in managing our emotional intelligence is that we do not have direct access to our unconscious processes, as we only know the result of our internal computations. We face the risk of inventing and confabulating reasons and explanations after the fact.

The Bias of Emotional Reasoning

The Emotional Tail Wags the Rational Dog

In some ways, emotions could drive rational thinking. Since our perception of reality is shaped by the way the world affects us, we tend to base our opinions and decisions first and foremost on our emotional perception. What we feel colors what we think. Our reasoning and beliefs are colored and filtered by our interoception and our affects. The emotional tail wags the rational dog, and our emotional attitude toward some big issue drives our mindsets and behaviors. In fact, our beliefs and political postures have been shaped by the many experiences we have been through, and this intense work of integration remains subconscious, implicit, and difficult to access. And when we want to justify our opinion, we often invent reasons that are meant to be rational, in the heat of the moment.

So, according to the *bias of emotional reasoning*, we follow our hearts and what we like determines what we value. Everything we feel to be good is good; everything we feel to be bad is bad. Donald Trump's enthusiasm for the potential of hydroxychloroquine as a treatment for coronavirus resulted from a 'gut feeling', as he puts it.[5] As he felt this was a good treatment, he recommended it. The same emotional reasoning also applies after a failure. The fact that 'I failed and was not up to the task" leads to the feeling that "I am worthless, I am a loser!".

[5] "I feel good about it. That's all it is, just a feeling, you know, smart guy. I feel good about it", declared President Trump at a press conference on 20 March 2020, about the use of hydroxychloroquine as a Covid-19 treatment (https://www.statnews.com/2020/03/20/trump-coronavirus-drug-just-a-feeling/).

Moreover, when emotions swamp the stage of the working memory, there is less space for rational evaluation. Fear can seriously disrupt the thinking brain. Thus, leaders should be aware of the fact that fear, anger, or the thrill of success, could greatly cloud their thinking.

The Political Brain

Drew Westen studied how people decide whom to vote for in presidential elections and observed that their reasoning was *motivated*, that is colored by an emotional investment. The political brain is not a dispassionate, calculating machine as voters think with their guts and not with economic data, or policy statements. They mainly pay attention to arguments that engender their enthusiasm, fear, anger, or contempt. Voters remember less what you told them, than the emotions they felt. Their reasoning is largely motivated by their feelings, and they want to sense an emotional resonance with their leaders.

When Donald Trump talks to his audience, he does not burden himself with numerical proof or verified data. He prefers to evoke the good old days, an idealized vision of America, and does not hesitate to stir up fear by conjuring up threatening dark forces.

Court Decisions

Several studies show that court decisions can be influenced by the emotional state of the judge. For instance, a suspect could either be released on probation, or remain in prison, depending on whether the judge has an empty stomach, is tired or has a bad first impression of the suspect. This emotional trigger may be sufficient to unconsciously influence the judge's decision. To avoid this bias, it is necessary to rely in parallel on facts and objective elements.

The Fear Circuit

The two main emotional mechanisms put in place by evolution to help us react and rapidly adapt, are the fear circuit and the reward circuit. We consider them successively.

The Fear Circuit and the Amygdala, the Emotional Sentinel

If we were to push our thumb far into our brain just above our ears, we would find an area called the *amygdala* in the singular, although we have two, located deep within the temporal lobes, left and right. The amygdala (*almond* in Greek, because of its shape) is the main center of the *brain fear circuit*. As an alert and vigilance hub, it performs emotional surveillance of our world, gauging the emotional significance of everything we see and deciding whether it is worth paying attention to it, even before the thing reaches our conscious thought. With its connections to our senses and all other regions of the brain, the amygdala orchestrates all our emotional experiences, as well as trust or mistrust in social relationships (see annex 4).

Bias of Fear and Insecurity

Fear speaks to our most primitive concerns. It has emerged from evolution to solve the recurring problem of survival in a dangerous environment and is now inscribed in our minds as a bias, a predisposition that make us anxious of detecting and anticipating any danger, as we face a multitude of incidents and problems all day long. Fear, alertness, anxiety, anguish, fright, dread, panic, terror, this long list reveals the importance of this bias in our daily life and in several mental disorders, notably phobias. The result is stress and sometime magical thinking as people often call upon fortune tellers or astrologers when distressed. The human brain is wired to overestimate and overreact to risk and threat before anything else. Distrust is more natural than trust.

Fear remains pervasive in organizations, creating a climate of insecurity. Few managers readily accept their staff making mistakes, even if they often profess the opposite view. Thus, errors, when they occur, are systematically hidden. In the same way people refrain from speaking up in a threatening environment established by their hierarchy. As a result, the fear of making errors and the aversion to questioning the status quo paralyzes the learning process and inhibits risk-taking.

We constantly need to control our environment and seek reassurance to avoid unpleasant surprises. The sight of a stranger's face triggers the amygdala within 40 ms to check whether it is a friend or a foe and activates a fight-or-flight response in consequence. It then takes almost half a second for the information to become conscious and correct the automatic reaction if necessary.

When our emotions become too intense faced with a serious threat (burglary, car accident, our boss shouting at us, or simply giving a speech), we are under stress and the amygdala has the power to override the prefrontal lobes. The more we feel emotional, the less we can access our rational mind. Within a split second, the amygdala can overwhelm our prefrontal cortex and hijack our ability to act reasonably. It then takes a few minutes to flush the chemicals released by emotions out of the bloodstream and calm down.

Cooling Down Emotions

At their current stage of evolution, the cognitive and the emotional system of the brain are not well connected. The connections between the amygdala and the frontal cortex are not symmetrical. They are much stronger from the amygdala to the frontal cortex than vice versa. The result is that the ability of the amygdala to control the cortex is much greater than the ability of the cortex to control the amygdala. Metaphorically speaking, we use country roads to calm down our emotional brain, but the emotional brain uses superhighways to flood our consciousness. Moreover, the braking system that inhibits the action of the amygdala functions like a muscle that gets easily tired.

Another way to manage emotions consists of shifting to a cognitive mode for the appraisal of the situation. The idea is to step back and *reappraise* the situation to calm the emotional response before it becomes unmanageable. What makes us emotional is not the situation itself but the way we think about it, the way we represent it, the way we interpret and frame it in our working memory. This implies (referring to the theater metaphor) that we can quickly change the decor and lighting of the stage, that is, change the context and the way we frame the situation. By taking another perspective, we can experience the situation differently.

Imagine, for example, that as you drive to work, you are stuck in a traffic jam. Soon you become bothered, impatient, or anxious. But you can avoid being driven into this emotional loop if, as soon as you notice the difficulty, you decide to reframe your appraisal and find three or four advantages of the situation: "Let's avoid a bad day and use this time to call a friend, listen to good music, or prepare for my next meeting". You only have a few seconds to change perspective because as soon as you are in the grip of an emotional takeover, your prefrontal cortex is overwhelmed by your stress. In fact, reappraisal is what we often do with the help of a coach or a therapist when we are not able to distance ourselves from a problematic situation.

The Reward Circuit

The Dopamine Reward Circuit

To survive, our ancestors had to anticipate and avoid dangers and threats, but they also needed to find new food sources and opportunities for reproduction. They relied on the brain's powerful system of pleasure seeking and immediate gratification, *the reward circuit*. The main hub of this circuit is the *ventral striatum*, referred to as the *rewarder*, that detects and anticipates rewards through the action of a neuromodulator, the *dopamine*. The rewarder represents a sort of craving button that sparks temptation and desire for immediate reward and plans specific actions to obtain it. Dopamine is released from a region located deep down in the brain stem, that we call the *sprinkler* and that broadcasts the sensory stimulus indicating a reward. The rewarder and the sprinkler are thus the main components of the dopamine reward circuit that elicit the

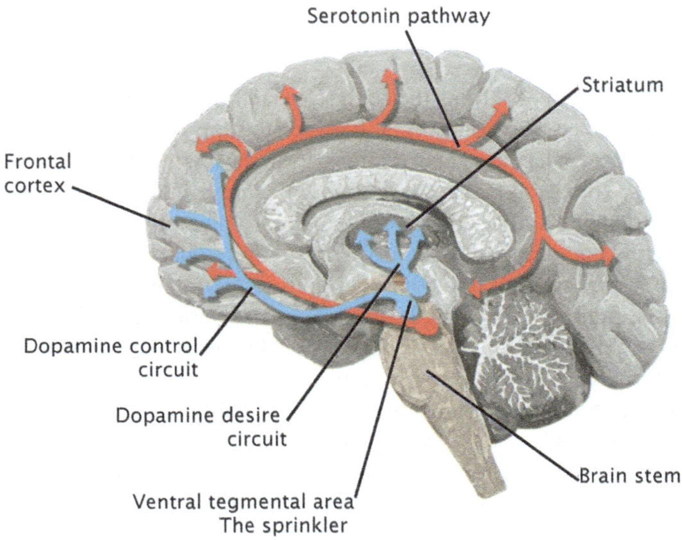

Fig. 2 Desire and control circuits. *The desire and control circuits are dopamine pathways. We have also represented another system, the serotonin pathway which is involved in arousal and mood. The neurons are clustered in the brain stem and extend their axons to several regions of the cortex, the ventral striatum and the amygdala*[7]

motivation and the wanting behavior. To simplify understanding we call it the *desire circuit*[6] (See Fig. 2 and annex 5).

Dopamine, the Stimulator

Dopamine is a powerful *stimulator* (hence the name we give it) which drives the quest for reward and novelty. It provides a rush of energy and pleasure when our goals and expectations are met. It reacts more from the anticipation of the reward than the satisfaction of obtaining it. Pleasure comes more from experiencing appetite than feeling satiated, more from the wanting than the liking. Wanting and liking are produced by two different systems in the brain.

Dopamine helps us accurately assess the difference between what is achieved and what is expected. The release of dopamine is stronger when the results achieved exceed expectations, or when we experience a pleasant surprise. This is a strong encouragement to carry on and pursue the action that led to such a reward. On the other hand, less dopamine is released in the opposite situation, as the dopamine

[6] Term borrowed from Lieberman and Long (2019).

[7] Low concentrations of serotonin seem to be associated with depression or aggression, and extremely low levels correlate with suicide attempts.

system is bidirectional. We then feel a *lack* or a *loss*, and this lack may propel us into motion. As a result, dopamine plays a key role in learning through reinforcement.

According to Plato, this feeling of lack characterizes desire. Once fulfilled, the need which fueled desire loses some of its strength, leading to what we call satiety. The same is true if the surprise of anticipation disappears and the reward becomes habitual. There is then little or no release of dopamine. The habituation process takes hold.[8]

The Bias of Immediate Reward

Dopamine neurons are thus the basis of a very persistent incentive system, which continually seeks to experience more pleasure and will put in the time and effort required to achieve it. It was important to have such a system in a not-so-old world where food sources were scarce and sexual partners few and far between. Thus, we are left now with a tendency to seek out more than what we experienced previously, always more, with the risk of addiction to immediate pleasure. This leads to what we call the bias of immediate reward akin to the sin of gluttony and not far from the sin of avarice, in the sense of greed and accumulation of wealth. This bias makes us want everything right away and can lead to shortsightedness and excess. A serious design flaw indeed!

The power of immediate gratification makes us easily push aside distant goals. For example, in the corporate world, many managers are willing to give up on long-term value-creating investments so as not to miss out on immediate profit. We tend, therefore, to be prisoners of the present as the consequences of our actions diminish or even fade in the future.

The Dopamine Control Circuit

When it becomes necessary to defer immediate rewards and introduce some thinking about future consequences of action, the urges of the desire circuit are opposed by another circuit, the *dopamine control circuit*,[9] that links the sprinkler to the prefrontal cortex through specific neural projections (see annex 5).

Now, the prefrontal cortex is involved in the control circuit to delay gratification, envision long-term consequences, and strategize. It is this ability to plan and produce these kinds of mental images about the future that makes us human. Although we may be tempted for a moment by a huge slice of cake, the prefrontal cortex can bring us back to reality and temper our desire to eat it. However, the activation of these

[8] The inevitability of habituation causes boredom. Arthur Schopenhauer saw boredom as an essential phenomenon. As lack leads to suffering, for him "Life swings like a pendulum backward and forward between pain and boredom". *The World as Will and Representation*, 1819.

[9] Term borrowed from Lieberman and Long (2019).

regions is costly in brain energy and more demanding than the activation of the desire circuit.

Delayed Gratification and Temporal Discounting

Indeed, the rewarder in the desire circuit remains focused on the attractiveness of immediate rewards and discounts the value of any delayed reward. And the further the expected rewards are postponed in time, the more value they lose. This is the "discounting" effect. The value of a dollar tomorrow is lower than its value today. The value of tomorrow's dollar must be reduced—"discounted"—in relation to what it is today, to take this loss into account. In finance, adjustment or discounting rates are applied to more precisely evaluate the revenues which will be obtained tomorrow from an investment made today.

In fact, this discounting effect has been magnified in our society of impatience and quick fixes. When it is easy to have everything immediately, the ability to wait fades and the conscious activation of the prefrontal cortex to curb immediate impulses becomes ever more demanding. This explains the difficulties of investing in the future, in matters of education, disease prevention, retirement, or the environment.

Self-Control and Willpower

Regulating Our Emotional States and Impulses

Self-control is the ability to keep a cool head and restrain or prevent some urge or desire; not getting angry when somebody is not respecting you, resisting the temptation to eat that extra piece of cake, putting on a brave face when interacting with an angry customer or taking a break before reacting impulsively. It's the dopamine control circuit which gets activated, and it takes a lot of cognitive effort to exercise willpower. But willpower is like a muscle which soon becomes fatigued with use and needs time to recover.

Most diets fail because the conscious forces of reason and willpower are simply not powerful enough to control drives and urges. For decades, smokers and drug users have received information about the dangers of addiction, and yet information programs alone are not sufficient to change behavior. The signals coming from the prefrontal cortex are less and less able to inhibit the neurons of the rewarder and are finally not heard at all. This mechanism is at the heart of addiction.

This ability to withstand immediate pleasure and thus defer short-term gain to obtain a higher gain later is illustrated by the famous marshmallow test carried out in the 1970s by Stanford University researcher Walter Mischel.[10]

Self-Control and Marshmallows

Around 1970, Walter Mischel at Stanford University launched a series of famous experiments in which he presented kids with a conflict between short-term gratification and delayed reward.

He sat a series of 4-year-old kids in a room and put a marshmallow on the table. He told them they could eat the marshmallow right away but, as he was going away for a short period, he added that if they waited until his return, he would give them two marshmallows.

In the videos of the experiment, we can see the children squirming, kicking, hiding their eyes, and banging their head on the table, trying not to eat the marshmallow. The kids who could wait several minutes possessed some impulse-control abilities and a certain level of self-confidence. They could find a way of reframing the situation and shifting their attention to maintain some control. They looked away from the temptation or were able to think about other things. This thinking ability is an aspect of emotional intelligence, the capacity to understand and regulate one's own feelings and desires.

But as soon as some kids engaged the "hot" networks in their brains, there was no way they could not eat the marshmallow. "It's hard for the control system to beat the desire system by willpower alone. Like a tired muscle, the control system soon wears down and caves in, but the desire system runs effortlessly and endlessly".

Years later, Mischel followed up to see what had happened to the children in his experiment. Those who had been able to wait did much better in school and had fewer behavioral problems. They generally succeeded at what they set out to do, and 30 years later they had much higher incomes. Since then, various other findings also point to self-control as one of the greatest assets a person can possess.

The Navy SEALs (Sea-Air-Land special operations force of the US Navy) are a good example of individuals who push themselves to the extreme limits of self-control. We can see in action the human qualities, the motivations, and the training necessary to overcome weaknesses of the will.

---

[10] Mischel (2015).

References

Hume, D. (1740). *A Treatise of Human Nature* (book 3, part 3, Section 4.3).
Lieberman, D., & Long, M. (2019). *The Molecule of More*. BenBella Books.
Mischel, W. (2015). *The Marshmallow Test: Understanding Self-control and How to Master it.* Gorgi books.

Chapter 5
The Rebellion

Employees Rebel

Rebellion Against the Rational Pyramidal Approach

The Taylorian bureaucratic model has contributed to the advent of the consumer society thanks to the enormous economies of scale generated by the division of labor, the automation of operations and the rationalization of work in a rigid pyramidal structure. But the waste of human and material resources and the inflexibility of the system led from the middle of the last century to a movement of resistance. It was not so much the labor movement that took these issues to heart, as the rank-and-file employees who rebelled. They wanted to become full human beings again, without leaving their brain—or more exactly their prefrontal cortex—at the factory gate. They wanted to regain some meaning and control over their work. Then, it was customers who in turn rebelled by refusing to have their individual needs ignored, and to be forced to pay for the waste and costs generated by such a system.

The first challenge came from the human relations movement, which believed that the intelligence and the motivations of employees could be put to good use. This led to a leadership approach associating leaders and followers (Chap. 1). Leaders now had to mobilize followers by acting on their needs, motivations, or aspirations. It was no longer acceptable to consider that workers were only guided by their primary predispositions and biases, that they were lazy by nature and had to be bound by rules, constraints, or punishments, that they were content to obey without taking initiatives, that they were only motivated by monetary rewards or bonuses. Leadership meant that employees were looking for meaningful work that offered them more autonomy and responsibility.

© The Author(s), under exclusive license to Springer Nature Singapore Pte Ltd. 2023
J. Teboul and P. Damier, *Neuroleadership*,
https://doi.org/10.1007/978-981-99-5122-2_5

New Experiments

As people became more and more disillusioned with the consumer society and started searching for personal achievement and increased autonomy, the human relations movement opened the way to new experiments driven by aspirations such as giving meaning and coherence to action, greater control and participation in decision-making, or better collaboration within teams.

The socio-technical approach focused on participative management within autonomous and multi-skilled teams. The idea was to move away from the narrow specialization of tasks where each operator was supposed to do a very small part of the job before sending it to the next stage without worrying about what would happen next.

For example, in the Volvo automotive factory in Kalmar, Sweden, autonomous teams were tested with some success. The assembly lines were organized into production islands organized around teams which had the power to elect their leader and manage schedules and procurement. The main idea was to reduce hierarchical dependence and specialization by broadening and enriching tasks. Because of their versatility, multi-skilled workers could help each other and directly manage the quality of their production. They could share their ideas and contribute to improvement and innovation. This kind of organization required a high investment in training but reduced the costs of coordination and quality control.

The fact that the socio-technical model was experimented with in the Scandinavian countries with some success could be explained by the open and egalitarian social culture that exists there, and by the involvement of trade unions in work organization issues. However, it proved difficult to maintain the motivation of operators given the repetitive nature of the operations. If you keep adding boring tasks to boring tasks to enrich a job, you get something even more boring and just more complicated. In fact, the success of such approaches could not be sustained because the experiments remained too partial and lacked overall coordination and customer focus.

Another very important wave that swept through the world of industry and services was the Total Quality Management movement pioneered in Japan by two eminent consultants, Edwards Deming and Joseph Juran. The process was initially triggered by highlighting the huge hidden costs of non-quality and the focus on the inspection of finished products instead of investing in prevention. The method requires learning from defects to prevent their recurrence. This can only be achieved by giving back workers the responsibility to correct and prevent defects upstream. The principle is simple, but the difficulty lies in this reversal of responsibility and organizational change. Some of the responsibilities that were assigned to white-collar specialists in quality or methods had to be given back to the operators on the production line.

However, these attempts remained limited and partial, as operations continued to be *pushed vertically* by central planning rather than being *pulled by customer demand.*

Customers Rebel

Focus on the Customer

At the societal level, attitudes were also changing. Consumers who had been provided with similar products and a certain level of conformity by mass production, gradually felt the need to express their originality, and differentiate themselves from the mass. This need was satisfied when they were able to access more customized products as companies were differentiating their products with the opening of markets, and the increased level of competition.

As the central planning at the top and the vertical push of objectives all the way down the pyramid could no longer respond quickly enough to the needs of customers, the whole organization had to orientate itself toward the end customer who would pull demand through the whole value chain, making cross-unit cooperation essential.

Toward the Lean Organization

The Toyota production system, also called the *lean production system*, is one of the best examples of a process *pulled* by end customer demand instead of being *pushed* vertically from central planning. Customer demand forces the operational units to interconnect horizontally from one to another. This leads to waste reduction, efficiency improvement, and better transversal cooperation. The need for top-down control fades away with fewer hierarchical levels.

Who is in charge now? It is the customer. The traditional organigram with its rectangular boxes stacked from top to bottom is now an outdated metaphor that only represents one aspect of how the organization should operate.

The shift from vertical control to transversal cooperation between units (Fig. 1) combines the search for ever lower costs, with the search for advantages that will increase the value perceived by the end customer.

Transversal Cooperation is not Natural

It should be noted that transversal cooperation between units is neither easy nor natural. Your counterpart in the next unit, the next silo, is most often a stranger with conflicting interests, but you depend on each other. To improve the relational engagement of both actors, it is necessary to create a context in which they have an interest in working together, and a climate of trust that clarifies the rules of the game.

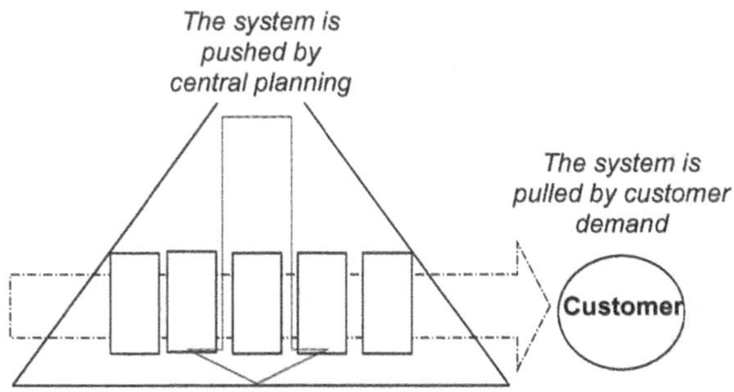

Fig. 1 Whole value chain is pulled by customer demand

Evolution Towards Services

The focus on the customer is further amplified when organizations provide services. According to our definition[1], services represent the sum of all interactions between *frontstage* employees and customers (the *backstage* dealing with material operations and transformations). This definition shows the crucial role of frontline personnel. They now have a real *relational power* as they control the relationship with the customer.

The best example that illustrates this definition is the restaurant where the service is delivered in the dining room, the *frontstage*, and the meal is prepared in the kitchen, the *backstage*. This clarifies the traditional opposition between two worlds, the world of services in the frontstage focusing on customer relationship and the world of production in the backstage focusing on efficiency.

The direct interaction between the customer and the employee (the diner and the waiter in a restaurant) is clearly represented on the *service triangle* as shown in Fig. 2. When the customer simply buys a commodity product, the relationship between the firm and the customer remains a simple transaction. However, when the sale of this product is accompanied by services, the firm must establish a relationship between the employee and the customer. This relationship empowers both protagonists by the very fact of their interaction. The firm therefore needs to select, train, and satisfy them simultaneously.

Unlike the sale of a product which can be reduced to a simple transaction when there are no added services, the delivery of a service requires that employees are given enough freedom of action to adjust to the needs of customers who are all different. These customers can then participate in the delivery of the service and *co-produce* it, while customizing it and reducing its cost. In the case of web platforms, for example, the customer coproduces the digital service by interacting on the interface.

Managers must now take into account the *relational power* that develops between their employees and customers and accept to perform a coaching role to support and help them deliver the best service. The organizational pyramid is turned upside

[1] Teboul (2006)

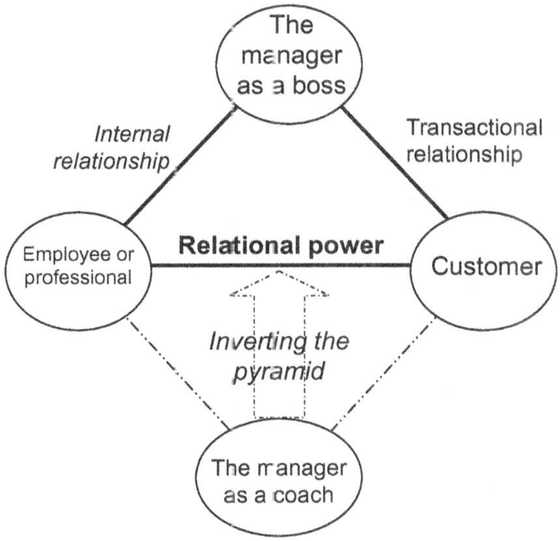

Fig. 2 The service triangle

down as shown in Fig. 2. This relational power, which is quite strong in the case of service professionals such as consultants, doctors, lawyers, or artists, often puts them in conflict with their managers.

Toward Transformational Leadership

Toward a Creative Model of Leadership to Mobilize Followers

Finally, to avoid facing rebellion from customers, employees, and other stakeholders, some leaders are tempted to leave the comfortable command-and-control approach to carry out a step-by-step transformation of the organization in association with followers. This means moving toward the creative leadership model that we have described in the first chapter, and which is organized along three main lines of action: an ambitious strategic vision, an open leadership style, and a favorable cultural context. They will need to discipline themselves and work hard against the natural predispositions and biases that we have noted but they also will have to motivate multiple actors and transform the organization into a *good place to work*. Will they have the resources? This is not an easy task when the intensification of competition demands the tightening up of managerial practices.

What Makes People Move and Engage?

What makes people excited about getting up in the morning? What makes people passionate about what they do? It is all about motivation and passion. Leaders must generate key motivators along the three lines of action.

Give followers meaning to their action when questioning and developing a strategic long-term vision.

Share the vision and empower actors to further explore and implement it with an engaged leadership style.

Recruit the right people prepared to collaborate and learn, and organize the cultural context to make it a great place to work in.

Main Motivating Factors

We have already mentioned the extrinsic motivators. They are external to the person and mainly represented by the metaphorical carrot-and-stick approach. Rewards and punishments operate right away in the fear system, or under the stimulation of dopamine in the desire circuit.

So, it is no wonder that salaries, bonuses, or similar advantages remain very powerful levers of action. According to Eric Schmidt[2] Google's executive chairman at the time, "The lower-level employee who helps create a breakthrough product or feature should be very handsomely rewarded. Pay outrageously good people outrageously well, regardless of their title or tenure. What counts is their impact".

Perks and other benefits have a similar effect. Google is renowned for the fabulous amenities available to their employees: free food, volleyball courts, subsidized massages, dog-friendly offices, bowling alleys, climbing walls and slides, gyms with personal trainers, and so on.

On the other hand, firing people and instilling a culture of control and fear would have an instant and disastrous demotivating effect.

Another extrinsic motivator is the aspiration to climb the hierarchical ladder to gain status and power. Again, the pleasure felt from a promotion will bring a surge of dopamine in the desire circuit, with the yearning for more.

By contrast, intrinsic motivators come from the interior of the person. People are moved forward with factors beyond money and status. The game is more complex as there is a range of options that have a more lasting effect and temper the attraction of immediate gratifications.

---

[2] Schmidt (2014).

Here is a quick review of the main intrinsic motivators:

Main Intrinsic Motivators

A search for meaning and achievement

People readily accept engagement if they can find purpose and interest in their work, if they can make an impact. They want their work to matter. Talented people are attracted to Google because they are empowered to make a difference and change the world.

A sense of freedom and autonomy

What motivates people is what they do and can achieve without constant monitoring. They want sufficient autonomy and freedom to achieve something meaningful. Their sense of accomplishment comes from mastering a difficult task.

A sense of enjoyment and discovery

Another way to motivate people is to give them the opportunity to explore new challenges and get fun from adventure. The pleasure of the trip and the joy of discovery is equal to its realization. It's about personal growth, learning, and professional development. The 20% time offered by Google, or the 15% time offered by 3M is about freedom, learning and collaboration. According to Eric Schmidt[3] "The most valuable result of 20% time isn't the products and features that get created, it's the things that people learn when they try something new. Most 20% projects require people to practice or develop skills outside of those they use on a day-to-day basis, often collaborating with colleagues they don't regularly work with".

A sense of being informed and in control

To accept to fully engage, people need reassurance. They want to understand what is going on and be informed with clarity about the direction chosen, financial stability, or the commitment of the leader. They want to be entitled to be in the loop and participate in the decision-making process. A sense of purpose emerges from a transparent and communicative environment and makes employees feel more in control overall.

A feeling of being included and recognized

People want to work with others who are as smart as they are and are interested in what they accomplish together. The social feeling of being included

[3] Schmidt (2014).

in a united and active community is a very basic need. So, good teamwork, contribution to a winning team, and identification with the group create an inspiring organizational culture.

Employees want to get consistent appreciation for their work. Recognition of accomplishments could be more crucial than financial rewards. People want to be noticed when they have done a good job. Constructive feedback is important.

A sense of fairness

People need to feel that everybody does their share and gets the same opportunities and credit that others do. Nobody takes advantage of the system.

Regression to the Fixed Model

When leaders do not or cannot make enough efforts to motivate and engage followers, they will tend to regress toward the fixed model given the natural predisposition of the brain for inertia and least effort. They will have difficulty maintaining a coherent approach, as biases will take over and dominate on one or the other of the three dimensions of the fixed model. The bias of immediate reward may lead back the strategic orientation toward short term objectives and results, the leadership style may show a lack of commitment and empowerment under the influence of the hierarchical bias, or the bias of insecurity will maintain a transactional culture that will not encourage informal trust and learning.

This will be more carefully analyzed in the following chapters, but we must give one more consideration to the fact that we are conditioned by habits and a predisposition to least effort.

Routines and Bias of Least Effort

When the workspace of our consciousness, the 'notepad' of our working memory, is saturated, we rely massively on automatisms, habits, and heuristics (rules of thumb) recorded in specific regions of the brain (see annex 7). This functioning is based on the search for the least effort using networks that are already wired in place.

Bias of Least Effort

The brain is a cognitive miser that seeks to perform the fewest mental operations possible for a satisfactory result. We prefer to operate according to an automatic mode of functioning by using recorded routines, habits, or empirical rules and go along well-trodden paths instead of relying on slow and deliberative thought. We go down the lines of steepest slope and least effort and are reluctant to use the laborious processes of rational reasoning. The desire to learn gives way to lazy thinking. This explains the attractiveness of default choices or status quo. This predisposition to expedience leads us, when we come to hesitate between what is easy and what is true, to choose, all too often, easiness.

We keep on sitting in the same spot, and we go for the first option when looking at the clauses of a contract. The power of inertia should never be underestimated, as some people may take advantage of it. When we choose to watch a TV channel, it is unlikely that we will turn over for the whole evening. Likewise, we rarely cancel the automatic renewal of a newspaper subscription. Equipment manufacturers always install in their devices a standard option that works by default and does not involve an active decision.

Our brain likes automatic thoughts, such as clichés and banal expressions that easily come to mind. It also gets carried away by emotional inferences, like emotional reasoning which makes one say, "If I feel it's fair and right, well, it must be".

In organizations, many people are content to obey, conform to the rules, and stick to the usual and reassuring way of doing things. If it ain't broke, don't fix it. People follow the path that already exists and is clearly indicated. Career progress is well delineated along hierarchical lines. Why seek to do more?

After this general overview, let's look now at the creative leadership model in action.

As the Western world was experimenting and developing new forms of leadership, a very successful management approach emerged in Japan at Toyota and spread in the rest of the world under the name of Toyotism and lean management. So, after theory and abstraction, let's now look at some concrete examples.

References

Schmidt, E. (2014). *How Google Works*. John Murray Press.
Teboul, J. (2006). *Service is front stage*. Palgrave Macmillan

Chapter 6
Creative Leadership in Action

Toyotism

The best way to illustrate the shift of perspective that can occur when leaders manage to mobilize the aspirations of their followers to improve the effectiveness of the organization is to look at the way Toyota reorganized the traditional manufacturing system into the famous *Toyota Production System* also known as the *lean manufacturing system*. This is the first illustration of the creative leadership model in action.

The Beginning of the Story

Toyota is one of the world's largest automotive groups and has established itself as the leader in hybrid vehicles. This Japanese company has remained consistently successful in the highly competitive automotive industry, thanks mainly to the development of a model of industrial organization known as TPS, which many companies have tried to copy, with varying degrees of success.

The story begins in the 1950s, when the company's leaders traveled to the USA to study American production methods on the Ford and General Motors production lines. After their visit, they realized that they had neither the means nor the luxury to afford such a waste of space, resources, excess stocks, quality problems, or hyperspecialization. Indeed, what they saw was the traditional Taylorian bureaucratic system which had not much evolved. They thus endeavored to set up a new system by applying the systematic improvement approach that they had originally used in the manufacture of looms.

© The Author(s), under exclusive license to Springer Nature Singapore Pte Ltd. 2023

J. Teboul and P. Damier, *Neuroleadership*,
https://doi.org/10.1007/978-981-99-5122-2_6

What They Saw Was the Traditional Taylorian System

One of the main things that caught their attention was the huge stocks that were sitting between different operations, for example, between stamping, pressing, or welding stations. Blue-collar workers that were assigned to carry out these operations had very narrowly defined tasks and were under strict supervision while white-collar workers proliferated, in functional, technical, or support roles. In this traditional command-and-control structure, decisions flowed down as only top executives had comprehensive information. Production lines were organized according to a central plan that defined the objectives that were pushed down from the top, with a heavy hierarchical structure of 6 or 7 layers.

A New Strategic Vision

From Pushing Objectives to Pulling the Flow

How did Toyota's executives address the accumulation of problems and sources of waste and complexity that they saw? The first thing to do was to come down to the shop floor and establish a new way of operating. Instead of *pushing* objectives from top to bottom, the new approach was to look closely at the flow of operations and see what was happening when this flow *was pulled* from final customer demand. The new imperative was to focus the whole organization toward the final customer as already mentioned in the previous chapter.

By pulling the flow horizontally from one unit to the next as if one were pulling a long rope connecting different units, everyone could see the waste, represented by idle loops between units. Pulling the rope would permit the elimination of the idle loops representing waste, one after the other. All actors on the production line could visualize the real situation by seeing for themselves what was happening on the shop floor and how things could be improved step by step.

By pulling the flow of operations from downstream, from the final customer demand, transversal cooperation between units and trades becomes inevitable. Contrary to the Taylorian approach where *central planning pushed* the objectives from the top to the bottom of the pyramid, Toyota *pulls* very gradually and *levels off* the flow stemming from the end demand to make problems emerge successively.

An Engaged Leadership Style

Sharing and Deploying the Vision

The common vision is now shared by all operators who closely follow the flow of operations and observe the sources of waste across units in concrete terms. The coordination between teams and various functions becomes more tangible and generates a stream of improvement projects to tackle all kinds of malfunctions and waste of inventory, nonquality, space, or delays.

Kaizen, the Continuous Improvement Approach

The principle of continuous improvement, known as *Kaizen,* can also be explained with the analogy of a river that does not run smoothly because its flow is hampered by boulders under its surface. If the water level is high enough, boulders are hidden. This means by analogy that, when the level of stocks is sufficiently high, problems are hidden. Now, when the water level in the river is lowered, i.e., the level of stocks is reduced, the first boulder, i.e., the first problem appears and can be treated. Then, as the water level continues to be lowered, and the level of stock continues to be reduced, more problems become visible, and things improve step by step.

Let's take, as an example, the stock of parts that existed between the presses and the welding operations. The huge stock there was the result of the fact that it used to take about 8 h to change the tool in the press (complex operation and heavy components). The consequence was that you had to produce long series of the same part to compensate for this long set-up time. So, the solution was now obvious. The set-up time had to be reduced drastically from 8 h to one minute. This major innovation, unthinkable in the Taylorian system of separate trades, was undertaken, step by step, with relentless determination, over a period of several years. At the end of this transformation, the press was able to produce very small series and the buffer stock was considerably reduced. This challenge forced specialists to cooperate transversally. The solution looks obvious now: if it takes one minute to change the tool in the press, a stock level of a few parts is sufficient to balance the line and it becomes possible to operate almost just-in-time, with a more regular flow within less space and with less accounting oversight. The same approach could be generalized to all kinds of concrete and tangible measures such as wasted space, walking distances, unnecessary transport and travel, or excessive delays.

This orientation toward the final customer gives everyone a way to formulate concrete and meaningful goals and identify with the team and the whole organization. The step-by-step approach at the lowest level of operations allows Toyota to manage the consequences of local improvement without destabilizing the whole system. A new way of doing things is tested, and, if relevant, it becomes the new standard.

Local adjustments make the system flexible and able to successfully adapt to a rapidly changing environment.

Empowerment

Turning to the concrete side of field operations, leaders not only display and communicate the new vision with determination, but also deploy it by delegating greater authority to blue-collar workers. They no longer remain at the level of planning and control, but they go and see for themselves what is happening on the shop floor where they strive to develop a learning culture based on delegation and cooperation.

In the factory, managers have their offices near the production lines, which allows them to interact directly with operators in the event of problems or difficulties. They can access the most relevant information at the most basic level of detail to quickly find a solution. This way they avoid aggregating the information into abstract reports.

Operators are now empowered to set up their machines and carry out basic maintenance; they are responsible for the quality of their operations and can take initiatives, as evidenced by the cords that hang from the ceiling on the assembly lines, which Toyota calls "*andon*". The operators can pull these cords every time they notice an anomaly at their workstation and need help. And help will come very fast because the whole production line will stop if the problem is not resolved quickly. These cords, which have been replaced by more sophisticated tracking systems in today's factories, function as alarm signals, sufficiently visible and prominent to encourage a quick and definitive resolution of the issue.

Shaping the Context

Hiring Policy

As empowerment, horizontal cooperation and continuous improvement became core principles of organization, the company made a considerable effort to hire and develop employees capable of adapting to a teamwork culture, and who were ready to experiment and learn. Contrary to the traditional Taylorian system, employees were hired with a very good level of education which they continued to develop through multiple internal training courses. The company therefore seeks to retain them and recruits its executives from its own ranks to perpetuate the culture. For Toyota, human resources represent a key social capital. It should be noted that the company could rely from the start on the Japanese social culture and the fact that unions strongly identify with the company. In this culture, strong team spirit and intense internal communication promote teamwork and mutual adjustment to reach consensus. Moreover, the Japanese have long been recognized as hard workers, eager to learn.

Flattening the Pyramid and Reducing Complexity

As managers work closer to field operations and teamwork is prevalent, the number of hierarchical levels is reduced as well as the complexity of interactions. Fewer procedures are necessary because operators are now responsible for what they do with much less supervision: machine set-up, maintenance, and quality among other tasks. Moreover, the communication between different units is made direct and unambiguous. Interactions always occur with the same, well-known person, analogous to runners passing the baton in races. This allows people to know their counterparts well and develop a close relationship with them. Similarly, each product or service follows a simple, direct path. It does not go to the next available person or machine, but to a specific person or machine.

Making Hidden Problems Visible

A key organizational principle was to multiply visual controls to *make hidden problems visible*. The fear of being at fault or guilty drives most people to hide their mistakes. But without mistakes, no learning is possible. Making problems and deviations visible is accepting learning from errors. Detecting an error at the source leads to immediate action. According to the same logic, machines and equipment are designed to stop automatically as soon as a problem arises.

Specific Methodologies

Several methods were developed to solve problems and improve operations. For example, the 5S method was based on the idea that to make things visible you must clean, scrub, and buff. A simplistic approach at first glance, but incredibly effective.

The 5S method in five steps

Classify. Keep only what is needed and get rid of what is not.

Order. A place for everything and everything in its place.

Make it visible. Expose what is abnormal and anticipate what can break.

Memorize. Create rules and standards to maintain acquired knowledge.

Persist. Discipline and habits to consolidate gains and prepare next improvement.

Standardizing

Creating a standard is a way of keeping a record of the way things are done. This helps define the starting point before taking the next step. This responsibility belongs to the operators doing the job. They write and apply the procedures themselves, progressing with the help of their supervisors, who guide them and teach them the philosophy of continuous improvement.

The new norm represents the memory of the present situation, the new record erasing the old one. Procedures must therefore be simple and practical enough to be used, changed, and updated by those who do the work daily. An operator who encounters a problem must immediately react and ask for support. If the operator is left to decide when to call for help, the problem may get bigger before it is solved, and the information on the real cause of the problem may disappear. Thus, managers cooperate with operators to improve the system and set the new norms. They act as trainers and facilitators by implementing problem-solving methods at all levels, including the lowest ones. Blaming an employee for an error or a problem does not solve it because it is the system that needs to be changed.

Reinforcing the Culture and Orchestrating the Whole

Implementing a Culture of Collaboration and Learning

Continuous improvement and learning imply the development of a strong culture of relational engagement in a climate of trust. This collaboration gives operators the opportunity to question and update their own operations with immediate feedback. Making an error is not considered a mistake. On the contrary, correcting errors is a source of learning and progress. In a way, it is not making an error that would be sanctioned, but the fact of not declaring it.

It should be noted that the lean organization that we have described reduces the distance between employees. Traditionally when people operate in a transactional mode, they are keen to maintain some distance while respecting their reciprocal territories in a polite and formal relationship. But when the distance is reduced, they must make a special effort to adjust to their counterpart and regularly renegotiate their relationship which is not necessarily friendly.

Orchestrating the Training and Sharing the Philosophy

A management center is essential to orchestrate the coherence of the whole. Most of the consultants involved in this center have received intensive training in the Toyota division specially created to develop and disseminate the management system throughout the organization. Many executives and senior managers have developed their skills in this center. They are often released from their hierarchical responsibilities to lead improvement and training projects in-house and with suppliers.

All employees receive multiple training courses that cascade from senior executives to first-level team leaders. Teams of consultants are available to deal with complex issues and to help senior managers reorient their units in line with the philosophy advocated.

Extending the Strategic Thinking

Many competitors who were quick to notice Toyota's success sought to copy its methods. But copying and imitating some practices is insufficient, because it is the entire system that must be adopted.

In fact, Toyota has pursued the same approach to address the main challenges which are now facing the automotive industry: supply management, innovation, and differentiation through services.

Concerning the supply chain, the TPS approach was extended to the first-tier suppliers. This led to a drastic reduction of their number and the development of a solid partnership with them.

Regarding the innovation aspect, the design and development of new products no longer occurs sequentially, but via teams working in parallel with a very high level of cooperation. Thanks to this collaborative approach, the launch time of a new model was reduced from five years to less than three years. The traditional sequential mode, where a project went from department to department, or, in fact, from one silo to the next, with validity tests at each stage, has been abandoned in favor of a multidisciplinary approach, whereby different departments remain in interaction throughout the project, developing a common culture. As strategic thinking is now focusing on the development of electric cars, Toyota seems to have taken a decisive turn, by being the first to successfully introduce hybrid technology.

Innovation also extends to services, to escape the hell of commoditization and develop a bespoke relationship with the customer, as nothing looks more like a car than another car. Toyota has made, and is making, a huge effort on the quality of services offered at the point of purchase and throughout the product life cycle, maintaining a solid relationship with the customer at each interaction point to build loyalty.

Turning Around Virginia Mason

The Toyota case allowed us to highlight the three main aspects of a successful creative model of leadership in action: the development and sharing of a strategic vision throughout the whole organization; the adoption of a leadership style that empowered all actors; and the establishment of an appropriate context and culture to mobilize them. This approach has inspired several other organizations, and we propose now to look at an illustration from a completely different domain. Let's see how Gary Kaplan succeeded in turning around the problematic situation at the Virginia Mason Hospital Center.

A Change-or-Die Situation[1]

Virginia Mason Medical Center is in Seattle (USA). This 350-bed medical center, with over 400 doctors and 5000 employees in 9 locations, faced a major crisis in the early 2000s. This crisis was the result of a loss of public confidence in hospitals. Within the institutions themselves there was also a drop in motivation and staff satisfaction, going hand in hand with less autonomy and greater management complexity.

As Gary Kaplan, the CEO, summed it up, "We don't have a choice. We change, or we die. It is that simple". And of course, the pressure of competition forced change in conditions that were not the most favorable. Hospitals are complex organizations to manage because of the differences in orientation and motivation of the multiple players and types of staff: management, board of directors, nurses and doctors, technicians, and researchers. As might be expected, the organization's short-sightedness, in the face of conflicts of interest from these various stakeholders and the pressure of short-term results, prevented it from anticipating and resolving problems before they resulted in significant cost overruns, lost revenue, or dramatic accidents.

Gary Kaplan did not want to follow the classic approach of the mission-obsessed leader, who promptly turns things around with an ambitious plan in search of quick and visible results throughout the organization. So, he set out to find an effective long-term approach.

[1] Bohmer (2010), Plsek (2013).

Trying the Total Quality Approach

At first, Gary Kaplan embraced the Total Quality approach of doing things right from the outset, with prevention and transversal cooperation between units. But the early trials were inconclusive.

Too much attention was paid to defects and errors, and in medicine, when an error occurs, the first question that arises is not "What happened?", but "Whose fault was it?", according to the expected *attribution bias*.

The same was true for formal process measurement and good practice which tended to go up the hierarchy, ultimately imposing a normative approach.

Gary Kaplan understood then that a common vision was needed to engage all stakeholders, from the board of directors to doctors and nurses, not to mention managers and executives. The whole organization needed to foster a culture of collaboration and learning. So, when Gary Kaplan discovered TPS, the Toyota Production System, during a trip, he decided to explore this approach that had to be introduced with determination and discipline over the long term.

The Strategic Vision

Focusing the Whole Organization on Patients

Following the 'change or die' message, a decisive staff meeting in October 2000 defined the terms of a new internal contract, mainly concerning physicians. Meetings followed to define a unifying vision for the organization, and discussions led to the decision to focus the whole organization on patients, now the top of the strategic plan pyramid (Fig. 1). This was not a foregone conclusion, as cutting-edge research and high-level publications were at least as important, not to mention the motivations of other players who continued to stir up debates.

This vision of driving the whole organization according to patient's needs meant that the focus was now on the patient's experience and the medical care while following the flow of patients across the organization.

Agreeing on a New Contract

During discussions, Gary Kaplan emphasized that customer orientation should not prevent the development of competences and expertise, as time and resources would be freed up with the reduction of nonvalue activities, waste, and lost time. What helped the final decision was the approval of the board members. They had been impressed and convinced by the results obtained by Toyota during a specially organized visit to Japan.

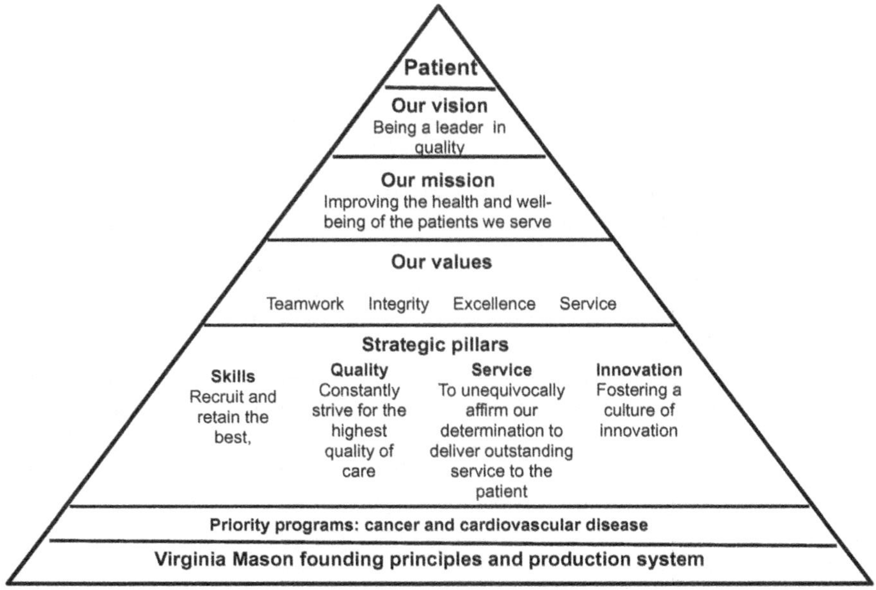

Fig. 1 The Strategic plan pyramid

So, Gary Kaplan finally managed to reach agreement on a new contract that would align the different players in a clear vision throughout the whole organization. Patient safety and experience would be the primary objective of the transformation.

The Strategic Plan Pyramid

Almost every presentation at Virginia Mason begins with the same image, the pyramid of the strategic plan, which fits on a single page, in a graphic form, fostering and deepening the consensus on the health and well-being of patients and specifying the strategic pillars (Fig. 1). Summarizing the essence of the strategy on a single page forces different stakeholders to negotiate and emphasize priorities. It's a nice way to acknowledge the limited capacity of the brain's working memory.

Leadership Style and Governance

A Resolute and Deliberate Long-Term Commitment

Such change can only be conceived and deployed over a long period of time. It needs to be prepared and thought out deliberately, with the will to push the learning effort resolutely. While Gary Kaplan's commitment seemed undeniable, his future and the length of his contract depended on the board of directors who were to engage the organization over the long-term.

So, Gary Kaplan worked hard to forge close ties with the members of the board. He organized discovery trips to Japan with some members to study the TPS methodology on the spot. He wanted them to be really involved and "get their hands dirty" by actively participating in the work of the teams.

The Leader as a Role Model

Gary Kaplan deployed this new approach by remaining modest and adopting a style of leadership open to discussion, without seeking to impose his point of view. The strategy was concise and explicit enough to be understood and accepted by all and to serve as a guide for making decisions.

To be viewed as a role model, he set up a patient safety alert system based on some incidents. For example, it so happened that a patient, who was about to undergo surgery, reported discomfort and a lack of numbness in the surgical site after the first injection of a medication that was in several syringes. The operation was immediately stopped, and the patient was placed under observation. It turned out that the syringes contained an incorrect medication mix. Gary then arranged for all those responsible to participate immediately in the search for the cause of the problem and eliminate it. A dual monitoring system was put in place.

Empowerment and experimentation. The principle of continuous improvement

While sharing the common strategic direction, Gary Kaplan empowered people to take initiatives by increasing opportunities for team experimentation with rapid improvement workshops according to the *kaizen* or continuous improvement approach and with workshops that lasted a few weeks and focused more on innovation than on simple improvement. This transformation could not be imposed and would only succeed if all stakeholders, physicians, nursing staff, technicians, pharmacists and even the patients themselves, were involved from the start.

Patient Guide System

A simple example of *kaizen* was the installation of a single-length handrail attached to the wall to give the patient continuous support from the bed to the bathroom. This avoided the need for a walker and allowed the patient to move without seeking the assistance of a nurse.

Healthcare-Acquired Infections

Another example is the systematic work undertaken to reduce healthcare-acquired infections. After mapping the existing monitoring system, the team in charge of improving the situation was able to question basic assumptions, habits, and routines that led to the current state of infections. Many suggestions allowed for some progress, however this is a long-term process, as for example, the discipline of having all staff and visitors routinely wash or clean their hands takes time to be firmly established.

The two principles that guided operations at Toyota were systematically used: processes and systems were to be improved by following the patient journey, and problems were to be made visible and controllable at the operational level.

Organizing and Redesigning the Context

Recruiting and Retaining the Best

Once this agreement on the new contract was reached, all stakeholders were required to honor it and seek help and advice, if necessary, but most importantly, it required those who did not agree, to leave the organization. As Kaplan acknowledged: "I recognized that you have to say good-bye. You can't keep everyone happy".

Moreover, the Virginia Mason's strategic goal of attracting and retaining the best talent meant that newcomers had to accept and engage in the learning and innovation process. It is important to avoid recruiting candidates who risk undermining the cultural harmony.

Making Things Visible on the Patient Journey

The treatment journey of patients represented a flow that crossed different units or departments. By mapping this flow, it was possible to bring to light all sorts of waste: waiting time, distance traveled by people, or physical space.

Nurses in Rooms and Hallways

For example, some studies showed that nurses spent most of their time walking around their care unit and staying in the central nursing station. Indeed, they only spent about 35% of their time in direct contact with patients. Teamwork led to the conclusion that a nurse could reasonably handle any four patients in adjacent rooms, if she or he had at hand the right equipment, supplies, information and, possibly, the help of a more specialized person. After reorganization, nurses did most of their work in the patient's room or in the hallway just outside the door. Their time available for patient care rose to almost 90%, and the average number of steps walked was divided by 10.

Fool-Proofing with Visible Indicators

Another example is the use of visible indicators to prevent infections when inserting venous catheters. The procedure involved attaching colored tape to a catheter inserted under less-then-ideal conditions, in the emergency department. This tape told doctors and nurses to change those lines as soon as possible, once the patient was stable.

Patient Safety Alert

The patient safety alert system which was the medical equivalent of Toyota's 'andon' system, whereby an employee on the assembly line could pull a cord to request help, was generalized. All staff encountering a situation that could harm the patient could trigger an alert to immediately inform a manager, and to slow down or stop the treatment process. Whether the problem was a misdiagnosis or treatment error, it was overtly identified so that systemic causes could be analyzed and corrected. The main concern was to avoid blaming the people in charge but instead to focus on the process itself and correct it. By establishing a climate where everyone felt safe to speak up, loud and clear if necessary, it became easier to understand the complex interactions that caused errors and anticipate them at an early stage.

Redesigning the Emergency Department in 2009

Lean management and improvement efforts do not prevent innovation. For example, Virginia Mason took on the challenge of reducing the size of the emergency department to the size of a small conference room. The first thing one notices upon entering the new department is that the waiting area is well laid out, comfortable looking, and quite unlike other hospitals. Patients are greeted by a clinician, and then assessed, and rapidly assigned to one of three flows.

The first flow is for patients who do not need acute hospital care. They are treated by nurses and support staff according to well-established protocols and guidelines, in a well-defined space. They do not occupy a bed and receive express treatment before being rapidly discharged home. The second flow identifies patients who are not severely ill but cannot be immediately discharged. In all other hospitals these patients remain under observation, creating a bottleneck causing overcrowding and forcing severely ill patients to wait. Here, they go to an adjacent accelerated care unit where they are stabilized and referred to the appropriate unit.

The third flow is for those patients with complex and difficult to diagnose problems and who need the full attention of the emergency department specialists. They occupy the main part of the service with 17 beds available. This triaging results in more attention being given by specialists to the few patients who remain in the main part of the emergency department.

A Fluid Context and a Learning Culture

Flattening the Hierarchy and Teamwork

Gary Kaplan was convinced of the necessity of creating a context and a *culture of relational engagement* by developing team collaboration and transversal cooperation between groups and work units. The flattening of hierarchy occurs naturally when people work in teams. As they focus on operational processes, teams tackle visible and actionable opportunities to reduce defects, waste, bottlenecks, or low value-added activities, but they can also generate wider-spanning ideas for innovation. As a result, teamwork was encouraged by eliminating overly rigid procedures, and giving all members of the team a way to participate, regardless of their position.

Relational Engagement

The increasing amounts of experiments grew the *relational engagement* of all actors. Effective functioning of a hospital, where tasks are inherently complex and interdependent, is enhanced when people establish an open, trusting relationship with each other while fostering a culture of cooperation, in contrast to the transactional culture where everyone merely plays an assigned role. The involvement of physicians in the design and reorganization of work processes facilitated team collaboration and generated familiarity and trust.

Establishing a relationship of trust does not necessarily mean being nice to each other or looking for good working conditions. It is about building a *rich interpersonal relationship* to give a feeling of comfort, openness, and confidence.

A Learning Culture

A learning culture requires a climate of psychological safety to allow people to learn from their mistakes but also look at things from a different point of view and eventually unlearn, that is write over previous knowledge and weaken the corresponding brain network.

Orchestrating the Whole

The Drive Starts from the Top

As the goal was to establish a culture of collaboration and learning that would unfold over the entire organization, the initiative and the drive had to come from the top. Indeed, when some leaders happen to show initiative at lower levels, they are often prevented from proceeding further by superiors who are not in agreement, or by rigid procedures that limit their access to critical resources.

A Systematic and Deliberate Approach

The leadership team knew that cultural change would not work on a piecemeal basis, in a disconnected way. A systematic and deliberate approach was required to involve the whole organization, from the board of directors to the frontline in direct contact with patients. This approach needed to be supported by the way the work of staff was assessed and rewarded.

So, career progression at all levels was based as much on the ability to collaborate in teams and develop interpersonal relationships as on technical skills and achievements. That was the case of physicians who were salaried, as their evaluation considered the nonclinical component of their activity. Some projects could lead to drastic staff reductions in some units, but there were no layoffs, as people were redeployed.

Instead of the traditional annual performance review that aroused intense emotional episodes, Gary Kaplan instituted a continuous feedback system, situation-based, more transparent and conducted by direct managers.

A Promotion and Monitoring Office

Overall consistency required a central structure to monitor and coordinate the change process, and above all, to organize training and foster education for all, over the long term. Thus, very early on, the leadership team decided to set up a central office, the *kaizen* promotion office, to support and accelerate the initiative. This light structure, comprising three managers for each division, played a catalytic role in supporting the adoption of the new approach. Its role was to guide change, not direct it. Its responsibilities included training and the consolidation of the culture of innovation.

Training was done in a variety of ways, either through job rotations or through a long stay at the *kaizen* promotion office, enabling the new trainee to prepare for a new leadership position. But the bulk of education occurred during experimentation.

The four-day main teaching course was gradually replaced by a series of two-to-three-hour modules. For example, there was a two-hour introduction module, mandatory for all new hires, that covered the basic concepts of the new approach. These modules allowed everyone to work at their own pace and focus more deeply on one or two concepts or tools at a time, applying them to examples or projects drawn from their work environment. The conclusions of important projects were presented after a preparation of several weeks, to a panel of experts.

In addition to general education programs, leadership training occurred at several levels. The program, which covered all aspects of general education, was completed with a final certification process for administrative directors, vice-presidents, or department heads. Education would gradually move from an individual basis to a group exchange approach, and the certification allowed holders to teach and conduct workshops and events. Lastly, the '*Kaizen Fellowship*' provided an additional level of development for those who had completed the certification. The aim was to accelerate the implementation of change with highly capable leaders and ensure succession.

Twenty Years Later

Gary Kaplan is still there, twenty years later, as president and CEO of Virginia Mason. He has received numerous prestigious awards and distinctions, both in the USA and abroad, for his accomplishment. When asked in March 2020 what he had learned during those years and what advice he would give to other leaders in the health sector, he said,[2] "One of the things I've learned to appreciate is the critical importance of culture. Culture is so important to our ability to continue to improve and get better for our patients and team members. Being at one organization over a long period of time, I've had the privilege of watching the evolution of our culture but also to lead it.

What I've learned is how the behavior of our leaders and the power of leadership through their behaviors could set and evolve the culture—leveling the hierarchy, eliminating fear in the workplace. These create a fertile field for improvement, for

---

[2] https://www.beckershospitalreview.com/hospital-management-administration/lessons-from-the-longtime-ceo-of-A-100-year-old-health-system-on-staying-relevant-in-healthcare.html?tmpl=component&print=1&layout=default.

people being the best they can be, to be thinking and becoming aware of all the waste there is in our [healthcare] environment.

Healthcare leaders at all levels stay relevant by staying curious. That curiosity manifests itself in their visibility, their questioning and humble inquiries, their being out of their offices with teams, coaching, facilitating, tearing down barriers. Many of us as leaders cut our teeth or became leaders by solving problems. But I think what we've learned—and where I think we can have the most value as leaders—is in building capability and helping through facilitation and empowerment, helping frontline team members to redesign their care, to come up with the solutions that are most meaningful to those teams and the people they serve."

References

Bohmer, R. (2010). *Virginia Mason Medical Center (Abridged)*, case 610–055. Harvard Business School.

Plsek, P. E. (2013). *Accelerating Health Care Transformation with Lean and Innovation: The Virginia Mason Experience*. CRC Press.

Chapter 7
Strategic Thinking

Rational Thinking

Strategic Direction as a Source of Motivation

Strategic thinking is a key competence expected from leaders who set the direction to follow and question the status quo to anticipate change. Setting the direction is a long-term engagement that should be shared at every level of the organization to define goals and priorities and remain open to new opportunities all the way down to field operations. This direction is often set in the form of a mission statement such as 'putting a man on the moon' or a core message such as Disney's 'making people happy', a vision that will later be made explicit in the form of objectives and priorities. Similarly, Virginia Mason defined a unifying vision by focusing the whole organization on patients.

Goals and priorities should be able to mobilize followers by giving them the feeling that they can make an impact, that they can give a meaning to what they do, or that they are part of a community united by a common ambition. To avoid the risk of fixing mindsets on short-term and abstract issues, profitability and similar measurements should not be mentioned at this stage.

Wegmans, a grocery chain that has been named regularly on the Fortune list of 100 *best companies to work for*, gives employees full discretion to do their best "not to let any customer leave unhappy". This company does not have to mention profit or market share objectives as it is self-evident that taking care of the customer regardless of cost is a good ambition that will bring results.

Direction and mission statements will serve as a guide for further strategic decisions to make the organization stand out, either by developing convincing competitive advantages or by exploring new areas less subject to competition. Given the dynamics of a very evaluative environment and the complexity of the multiple decisions to be made, the process of bringing out new propositions can hardly be defined solely

© The Author(s), under exclusive license to Springer Nature Singapore Pte Ltd. 2023
J. Teboul and P. Damier, *Neuroleadership*,
https://doi.org/10.1007/978-981-99-5122-2_7

from the top of the organization. Actors interacting with different stakeholders, on the front line or on field operations, are often best placed to know what is going on and suggest opportunities for improvement and action. But, whatever the new propositions, they will be incorporated in the main direction that guides action.

Rational Demands on the Prefrontal Cortex

We have seen that strategic thinking is first and foremost a rational approach based on the comparison of the cost–benefits of different alternatives or on sophisticated methods of choice optimization. Options are compared by analyzing their strengths and weaknesses against external opportunities and threats. This analytical work broadens the strategic vision beyond the usual frameworks and requires a relatively long-term perspective.

As seen in Chap. 3, laborious phases of reflection place great demands on the prefrontal cortex and the narrow window of attention available to access information. As a result, we have to work hard to carefully select and analyze the crucial pieces of information that deserve center stage. This is a big job that the brain, lazy by nature, is often reluctant to do spontaneously, and often, consultants are called in as backup.

The reasoner (the reasoning center located mainly in the dorsolateral prefrontal cortex) consciously keeps in mind, in our working memory, the useful information that allows us to make a decision. This working memory has a very low storage capacity and relies, in real time, on the brain's other long-term memories to retrieve the information it needs. This implies that producing complex reasoning is costly in attentional energy. Our rational capacities are much lower than what we think. We can only do one thing at a time and multi-tasking is a myth. Conscious functioning is slow, controlled, and sequential. However, it remains quite flexible, because we can quickly switch from one task to another, or change the rule.

In contrast, automatic mental operations such as simple mental arithmetic, recognizing a face, or reading a text are fast, consume little power, and use multiple processors that operate in parallel subconsciously. This is the domain of automated processing and habits.

Fortunately, decision making and rational calculation can now go beyond brain limitations, via external media and memories, such as paper, blackboards, computers, or networks. This is the domain of *cognitive externalization.* The mass of data now available can be analyzed and summarized using powerful statistical methods or mathematical models such as factorial analysis or regression analysis, which will extract as much information as possible to guide the decision. We are entering the field of artificial intelligence. For example, doctors who were confused by the multitude of factors at play when predicting heart attacks were able to extract a few risk factors by using factorial analysis from the data of hundreds of cases. Those were used to develop a simple decision tree to facilitate diagnosis.

Rational methods are quite valid and effective, but they are laborious and only use a part of our brain. Nevertheless, we can use a less conscious part of the brain that deals with predictive intelligence.

Predictive Intelligence

When people are faced with simple decisions like buying an ordinary household appliance, they usually have no difficulty comparing options, and they end up satisfied with their decision. However, when the decision is more complex, such as buying a new car or a house, they must keep in mind a lot of information, and they often end up confused as their working memory is overwhelmed with the profusion of details. So, they should not fully trust their conscious judgment. They would often do better by basing their choice on the massive parallel processing that operates unconsciously and feeds *inferences* and *predictions* in the form of *intuition*. This is illustrated by research conducted by Ap Dijksterhuis.[1]

In a typical experiment, participants were asked to choose the best car among many versions with multiple options. There was an optimal choice that participants were supposed to find.

At the start of the experiment, participants were given time to study ample information about the pros and cons of each version. Then, when the time came to make the decision, Dijksterhuis allowed one group to study the information for another 3 min while he distracted an other group with a task that prevented them from consciously thinking about the decision to be made.

Contrary to expectations, it was the second group, the group who made their decisions without further conscious deliberation, that fared better. Given the large number of variables under consideration and the restricted conscious space available for deliberation, it appears that the additional three minutes of data analysis did not help the participants of the first group to choose better. They had a hard time consciously juggling so many variables. In contrast, the participants who had been distracted were able to better rely on their unconscious processing, thereby producing useful associations between all the information provided. Their judgment was, ultimately, more relevant.

Our nonconscious processing sorts out, evaluates, and averages information to provide *inferences* that guide the decision when the conscious mind is overburdened. These inferences work in somewhat the same way as the global evaluation scores of a restaurant guide. They shape our interpretation of the world and help us anticipate and foresee what will happen. This *predictive intelligence* or *intuitive thinking* allows for instant choices. These quick decisions were essential when it came to survival in ancient times. In ordinary life, intuition allows us to get by in most situations, without recourse to a high demand of attentional energy.

---

[1] Dijkterhuis et al. (2006).

Unconscious processes can do an especially good job when they have been trained by years of deliberation. In fact, the training done in the clear light of consciousness has gone underground into the roomy basement of the subconscious. When we sleep, the mass of data we have accumulated during the day is processed, associated, and tested in our brain.

A Repertoire of Underlying Patterns

By associating and averaging a large set of similar situations experienced in the past, our brain ends up recording and coding the information in the form of concise *schematic patterns*, which broadly summarize the essence of these situations. These patterns then serve as a reference when a new situation is encountered. They are not directly accessible to our conscious register, but they emerge in the form of intuitions and *insights* which lead to quick, and often effective decisions. For the economist Herbert Simon, intuition was simply *pattern recognition.*

When we feel that an agreement is established between a few situational clues and a reference pattern, we often get a "gut" feeling, a bodily signal, not necessarily clearly felt. This signal is then transmitted through the *insula* to the prefrontal cortex, which takes it into account for interpretation and decision making. It is thus not uncommon to have an instant "gut" feeling that a situation is going to turn out badly or that a leader, albeit convincing, is lying.

Recognition-Primed Decisions

Intuition is a rapid judgment process which mostly eludes us. It is automatic and predominantly associative and unconscious. We also refer to it as an insight, a vision that appears like lightning and suddenly shows the situation with clarity, providing a feeling of certainty. Television series constantly require new actors, and directors mainly use their intuition to select them. It usually only takes them a few seconds to find out if the person is the right one. A few words and gestures are enough.

Our many experiences enable us to accumulate a stock of generic patterns to rely on to better anticipate what to expect when we come across situations comparable to those experienced before. These generic patterns condense a large amount of data which would have been processed with difficulty and less rapidly in our working memory.

For example, high-level chess players do not analyze the board square by square or piece by piece to anticipate possible moves by their opponent. They recognize the arrangements of several pieces together into specific patterns, and they compare them to similar patterns they have memorized after many years of practice.

Many professions are based on this ability to recognize underlying patterns or models. This is what happens when a physician prescribes a treatment based on symptoms. If she considers that the first solution considered is not suitable, she will imagine another solution by performing a new mental simulation. The best diagnosticians have memorized a larger collection of typical models that allow them to decide faster with fewer clues. Similarly, a lawyer can rapidly find the appropriate defense strategy.

The intuitive system works automatically without effort and has an extraordinary capacity to produce hypotheses and answers to usual situations. But the validity of results depends on the stability of the situation. If the domain of expertise evolves rapidly, it would be risky to rely on one's intuition and trust previous experience.

Intuition and Rational Reasoning

Many managers base their decision on an implicit knowledge of the situation and trust their intuition to choose the first valid option to act upon. Depending on the results, they may confirm the option or resort to another. They are more reluctant to use rational analytical thinking because it requires greater cognitive effort, as is the case when it comes to optimizing a choice by comparing the cost–benefits of different alternatives. They prefer this trial-and-error intuitive functioning, and scoff at slow, recondite analyses.

Intuition Must Be Trained

Behind intuitive inspiration, there is quite a lot of work to prepare and digest past information. Acquiring expertise in complex tasks, be it sports, music, science, or fire-fighting, requires the accumulation of knowledge and skills through regular practice with case studies, simulations, or experiments, to test hypotheses.

Correcting Intuition with Rational Thinking

Caught up in the dynamics of the moment, many leaders trust their intuition and make quick decisions by inference, drawing on their accumulated experience. But, most often, they do not consider the fact that their intuition is only valid in their specific area of expertise, and in a stable environment. And this applies equally to the advisors and experts they consult. The danger, then, is that they will embark on ambitious and heroic changes and restructurings that may get the attention of investors, but which are not very robust from a rational point of view. In the end, this may matter little to them, as they will have changed jobs before it would be possible to hold them responsible for the results.

Moreover, intuition presents a serious risk, as the hypotheses generated can easily lead to erroneous beliefs which can turn into solid convictions if they are not disproved by rational analysis. Rational thinking can be considered as an *inhibitory system* that regulates or corrects the assumptions made by intuition. Some scientific facts are counter-intuitive, and we must actively think in opposition to a part of our own brain. We have some difficulty in accepting that the earth is round, that the color white does not exist, or that most matter consists of emptiness.

Innovation and Creativity

The Fleeting Nature of Creativity

Intuition can lead to strokes of luck and creative leaps. Creativity can be defined simply as the capacity to make uncommon predictions via unlikely juxtapositions or analogies. As our concrete experiences grow richer, we allow our brains the freedom to search for as yet unknown connections. Signals from different areas of the cortex bounce around through all existing connections and patterns.

We can talk rationally about creativity, but creativity is not born out of reasoning. It develops out of the dance of patterns and chance encounters. It is about getting lucky when searching for new combinations in the wilderness of possibilities.

As the mind wanders without logical, conscious thinking, we may gain access to rich but evanescent representations. As soon as an answer is found, it emerges into consciousness. What led to the invention of the microwave oven were the unexpected discharges from a radar system that were melting candy bars in the shirt pockets of experimenters. George de Mestral invented Velcro, the interlocking of hooks and loops, after observing the burrs he found on his clothes after a hike.

Four Periods of Intellectual Creation

Preparation Period: A Long Search

Innovation often comes at a price as it involves thinking for long hours about the nature of the problem, elaborating on it, collecting all the knowledge available, and making connections among various pieces of information and experience. It is about doing the hard and often boring stuff, failing often, and starting all over again. But luck is earned through hard work. "It is the most elaborate form of skill", said Napoleon.

Incubation

Ideas are like plants, in that seeds do not look much like the final flower; they need time and care to survive and blossom. In the creative process, incubation involves setting up mental clues that are ready to fire. During this phase, it is important to have enough time and freedom to relax and let one's mind wander to let innovation occur.

Illumination and Insight

Then there is the classic moment of insight, the burst of inspiration, Archimedes' eureka moment, the splash of insight, the leap in learning, the cognitive snap, and the epiphany. A complex set of new connections is created as disparate observations integrate into a coherent whole, and the answer comes into focus all at once. Two things are important in this process: the reorganization of knowledge into new associative networks and the emotional impact of the discovery. Both effects will consolidate learning.

Verification: Refining and Implementing the Idea

Insight is not the end of the process. Engineers, designers, and others involved in taking a creative idea to fruition still usually have to spend years fine-tuning the unrefined idea, trying to make it real and concrete.

In Summary, Three Levels of Information Processing

Strategic thinking at the conscious level is supported by cognitive externalization and subconscious processing, as shown in Fig. 1.

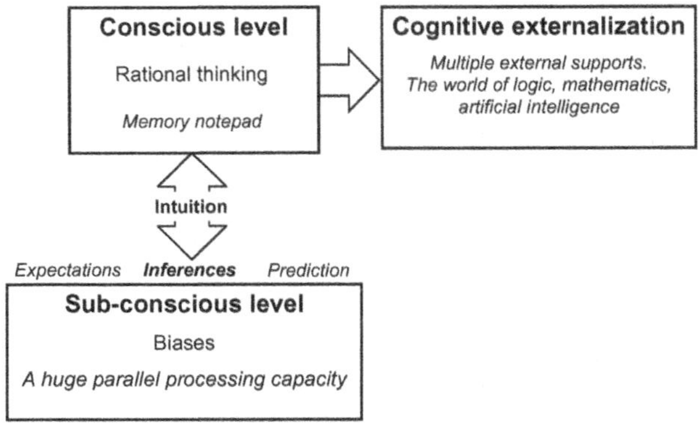

Memory notepad at the conscious level
Conscious attention has a very narrow space available in our working memory, only capable of displaying one thing at a time. Mental activity is limited by this "memory notepad".

Cognitive externalization
The advent of writing and external media has made it possible to offload the processing of reasoning and data analysis to external memories, thanks to mathematical or statistical languages and methods, massive data processing, and artificial intelligence algorithms.

Inferences and intuition
Multiple processors operate in parallel to associate, average and digest the mass of information coming from our past activities. They carry out a sort of statistical processing of this data to make inferences. The brain is a wonderful machine for foreseeing and anticipating the future.

Fig. 1 Three levels of information processing

Strategic Thinking is Constrained by Our Biases

The Bias of Immediate Reward Prevents Long-Term Strategic Thinking

Long-term strategic thinking must face shareholders' power as investors want immediate results that influence the share price rather than dividends. To attract financial analysts and investors, leaders must show sound quarterly results. These constant efforts at seduction end up making some companies dysfunctional or even corrupt, as evidenced by Enron's bankruptcy in the United States.[2]

[2] Enron was one of the largest companies in the energy industry. In December 2001 it plunged into bankruptcy, due to losses caused by speculative operations disguised as profits by accounting fraud.

The need to show convincing and immediate results is magnified by the importance attributed to finance departments in today's organizations. They reshape value chains by tightening bolts, outsourcing activities, and toughening up management practices. All this is done at the expense of the internal climate of the living organization and its future success. Long-term strategic thinking and stability can even be seen as a sign that the firm is not sufficiently innovative and dynamic.

As current results are overemphasized, future gains which could result from a more patient and persistent strategy are devalued. This operating mode is in line with the bias of immediate reward when the brain reward circuit gets activated by short-term stimulations.

Bias of immediate reward

We have already seen in Chap. 4 that dopamine neurons are the basis of a very persistent incentive system, which continually seeks to experience more pleasure and that immediate gratification makes us easily push aside distant goals. So, to avoid missing immediate profit many managers will give up on long-term value-creating investments. But when it becomes necessary to defer immediate rewards, the urges of the *desire circuit* are opposed by another circuit, the *dopamine control circuit* which involves the prefrontal cortex to delay gratification and envision long-term goals. It is this ability to produce these kinds of mental images about the future that makes us human. However, the activation of these regions is more demanding than the activation of immediate reward systems.

Indeed, the rewarder of the desire circuit, being focused on the attractiveness of immediate rewards, discounts the value of the delayed reward, all the more when expected results are postponed in time. This discounting effect is in fact magnified in our society of impatience and quick fixes.

In the present volatile and changing environment, short-termism pushes organizations to change rapidly, and after a restructuring, employees usually have no idea what will happen to them as this type of change is typically driven by debt and equity market imperatives, rather than by the internal dynamics of the firm over the long term. This unstable environment and the latent insecurity demotivate employees as they can no longer expect a coherent future.

The Confirmation Bias Prevents Questioning and Innovating

Strategic questioning is essential to maintain a persistent vision over time and anticipate change. But this questioning is hindered by an important predisposition of the brain. The *confirmation bias* plays an important role in reinforcing beliefs and making questioning, unlearning, and change more difficult.

Confirmation bias

What we learn is inscribed in our brains in the form of neural networks, and any new information is associated and bound to the neural circuits already present. Learning is thus done by integrating new information into the networks in place, confirming what already exists. Thus, confirmation bias reinforces our beliefs and tends to ignore or reject contradictory information. This explains why plans and objectives are rarely questioned. If the new information is consistent with previous beliefs, then it is easily integrated, and the belief is confirmed. If, on the contrary, it is at odds with previous beliefs, association becomes difficult, and the new information is easily rejected, modified, or forgotten.

So, our beliefs and values result from the way our lived experiences are averaged, summarized, and associated to previously stored information. They move deeper into our subconscious and form our identity. Behind the scenes, they guide our actions and behaviors and constantly give meaning to what we perceive or memorize. We readily trust our beliefs from the start and tend to confirm them before testing them in a rational way. The confirmation bias partly explains the success of fake news on the internet, as we easily accept new information that fits our preconceived ideas without testing it and we tend to confirm and reinforce it with all kinds of messages or statements in search of approval.

Framing

Confirmation bias also applies to the way we perceive, interpret, and *frame* a situation. Given our narrow focus of awareness, a situation can only be viewed from a specific point of view, according to the focus of attention. For example, a first impression or a particularly noticeable element, like physical appearance, gender or race, can have an excessive influence according to an anchoring or halo effect. This framing or specific perspective does not allow to see the whole view. Some aspects are reinforced while others remain in the shadows. Questioning is changing the framing by taking a helicopter view or moving one step aside. This is costly in terms of rational thinking and effort.

Scientific reasoning relies largely on inductive reasoning that in turn relies on multiple experiments to confirm that the same result is obtained each time. The result is that theories derived from inductive reasoning remain beliefs. For example, we may believe and theorize that all swans are white if all the observations so far recorded indicate that they all are white. But it is enough to observe a single black swan for the theory to collapse. A rigorous scientific approach tests the limits of inductive reasoning by organizing experiments that will refute the hypothesis and prove the theory false.

Confirmation Bias is Comforting

Our brain is comforted with a burst of dopamine, a burst of pleasure when our beliefs are confirmed. According to Bertrand Russell "What people really want is not knowledge but certainty". Hence, the reward from the confirmation of our beliefs can become addictive. The confirmation bias keeps us on well-trodden paths, making it difficult to unlearn and change. By contrast, we are destabilized by something that refutes our beliefs.

Cognitive Dissonance

Information that confronts and challenges our beliefs, or even more, that proves us wrong, leads to a painful internal conflict called *cognitive dissonance*. It is difficult to hold two incompatible or contradictory views or beliefs together in our minds. When we are confronted with potentially dissonant information that contradicts our beliefs, the brain tries to shut down distress by using faulty reasoning or by ignoring the contradiction. Moreover, it manages to activate the reward circuit to reinforce positively biased reasoning, with a rush of dopamine in the rewarder. We are ready to transform reality with all kinds of rationalizations or imaginary evidence to reach a reassuring interpretation that will preserve our beliefs. We distort and rationalize threatening evidence to feel good and avoid what is painful. It is a sort of defense mechanism.

It is not always to our advantage to see the world as it is. A heavy drinker perusing the latest statistics on the damage caused by excessive alcohol consumption will reduce dissonance by dismissing statistics and by convincing themselves that they are just casual drinkers, and, therefore, not directly concerned.

Willful Blindness

We also silence dissonance by closing the narrow door of awareness. Dissonance does not exist if we are not consciously aware of it. This is especially the case with devout believers or political experts who selectively interpret data so that they prove them right. Fundamentalists and extremists have such a huge psychological investment in their ideals and beliefs that they become impermeable to evidence that contradicts their group mythology.

In conclusion, taking the long path toward the creative leadership model requires mitigating biases and motivating people with long-term strategic thinking. Clear lines of action and priorities will mobilize the whole organization by giving everyone meaning and opportunities to make an impact.

Reference

Dijksterhuis, A., Bos, M., Nordgren, L., & van Baaren, R. (2006). On making the right choice: The deliberation-without-attention effect. *Science, 311*, 1005–1007.

Chapter 8
Leadership Style

The Leader's Personal Commitment

After setting the direction to follow, leaders must now deploy it and show their personal engagement to convince employees not only to follow them, but also to take on a leadership role. The organization will move from the traditional leader–follower mode to a more flexible and creative leader–leader mode. This leadership style involves three main aspects: first, the long-term personal commitment of leaders; second, their ability to communicate and share the chosen direction and goals; and third, their willingness to engage employees to cooperate, learn, and innovate, by empowering them.

Caring for Others and Long-Term Commitment

So, the first question that employees have in their minds before engaging further concerns the level of conviction and determination of their superiors. Can they be trusted? Are they driven by a sense of mission, or by their personal interest and a quest for power and status? Are they working for the health and survival of the organization, or for themselves?

In theory, leaders are supposed to take care of the interests of their staff before their own interests. Richard Branson, the famous British entrepreneur known for the success of the Virgin brand, shows his engagement by putting his employees first, customers second, and shareholders third. These are good intentions indeed, but will they last?

According to Jim Collins,[1] the successful transformation of an organization does not require charismatic or big personalities who make headlines and usually manage quick turnarounds by slashing the workforce and costs, reducing the R&D budget,

[1] James Collins. *Good to Great: Why Some Companies Make the Leap... and Others Don't*. 2001.

© The Author(s), under exclusive license to Springer Nature Singapore Pte Ltd. 2023

J. Teboul and P. Damier, *Neuroleadership*,
https://doi.org/10.1007/978-981-99-5122-2_8

and focusing on short-term results. This is not to say that charisma and bright personal style are not important, they can help, but contrary to expectations, leaders of great organizations are often unpretentious and modest and focus their attention on the leadership process. They talk about the contributions of other executives and minimize their own role. In times of failure, they take the responsibility. But they show an intense and unwavering professional resolve while refusing mediocrity in any form. They first and foremost think about the success and the future of their organization.

So, faced with stress, administrative and budgetary negotiations, or market uncertainty, leaders, at all levels, must show determination and resilience to face the unexpected. Their strength of character is their strength of conviction. By focusing on limited goals and priorities, and persistently pushing in the same direction, they should be able to overcome obstacles. This is in line with the '*hedgehog concept*'[2] which according to Jim Collins represents the single big vision that great leaders pursue with perseverance over time.

Leading by Example

As they lead the way and share their vision, leaders must be trusted. To believe the message, you have to trust the messenger. So, leaders will reinforce their credibility by setting an example in the way they behave and aligning their talk with action. In today's media-saturated world, they are bound to reveal ever more about themselves. Physical stature or style of interaction can help, but above all, they are expected to be a part of the story they tell and to move from speech to action. In the end, they should fuse with their strategy, like the dancer with the dance. To paraphrase Gandhi, they are the change they want to see.

This is not to say that leaders must always remain authentically themselves because this could put them in contradiction with the way they should act at a given moment. Some situations require a great deal of self-regulation and emotional stability to remain natural and credible as the circumstances require. In difficult situations it is essential to show courage and resilient optimism to confront risk and uncertainly. Who would want to follow a pessimistic leader? This brings us back to emotional intelligence, the ability to know oneself, to become aware of one's behavior and control its impact on others. It is hard to lead and motivate others if you are unable to manage yourself.

Thus, the creative leadership model requires determined leaders, committed over the long-term to build trust and engagement for all. They must also be able to communicate and share their vision with sufficient conviction to enable their followers to engage in the same way.

---

[2] Berlin (1953), A fox knows many things, but a hedgehog knows one big thing. Hedgehogs interpret everything in the light of some single all-embracing system, a single pattern that they believe in and adopt. Foxes have a more pluralistic approach and are able to take many directions.

The Limits of Good Resolutions and Egocentric Bias

When employees are questioned about the way they are managed, their first response is a complaint about how managers abuse power to protect their own interests and position. Studies in the United States have shown that the ratio between the remuneration of top managers and employees, which was of the order of 20 in 1965, forty years later had exceeded 200.

Of course, the world would be more human if leaders were genuine, modest, and concerned with the well-being of others. But self-centeredness and self-interest are predispositions that are deeply rooted in the brain and the real problem would be to deny it. We are inclined to make choices that serve us even though we think we are less self-centered than others. This tendency to put oneself first is fueled by the egocentric bias that puts our ego at the center of our attention, as already noted in Chap. 1.

Egocentric bias

Being at the center of our attention, we strengthen our self-esteem by overestimating who we are, what we think, what we know, what we own, or what we do. We are also attracted to anything that can boost that self-esteem. This may be professional success, compliments and applause, the quest for fame and adulation, or status. We give more credit to our own beliefs and creations than to similar ideas generated by others. Competitors are thus regularly underestimated. This excess of self-confidence limits our ability to truly listen and accept feedback and evaluations from others. This was the case of Decca who refused to sign the Beatles, or General John Sedgwick who lost his life at the Battle of Gettysburg ("At this distance they couldn't hit an elephant!" he claimed while refusing to seek shelter).

Inflating our self-esteem and cosseting our narcissism provide pleasure. For example, a promotion to a better position that boosts our status and self-image triggers an injection of dopamine, a spurt of happiness, into the rewarder. But when our status declines, such as when we receive criticism or derogatory remarks, the opposite happens: our dopamine level decreases, and we feel a lack of satisfaction. It is interesting to note that the 'likes' that pop up on our Facebook, or other social media profile pages, similarly reinforce our self-esteem. As dopamine is at work, we tend to seek more stimulation against the background of habituation, and this make us dependent and addicted. Then, beware of criticism and denigration, the minuses weigh much more than the pluses, and the damage can be considerable in terms of image and lack of self-confidence.

Testosterone contributes to the egocentric bias by increasing self-confidence and driving people to be dominant, arrogant, and overly optimistic. When they get to the top of the ladder and fight to maintain their status, a release of testosterone boosts their self-confidence. Up there, they have no more hesitation in taking credit for the work of others, while paying less and less attention to diverging opinions.

On average, men have five to ten times more testosterone than women, which explains why they are more prone to being self-confident. In fact, testosterone is a powerful inhibitor of oxytocin, a neuromediator that plays an important role in trusting others. Consequently, testosterone reduces empathetic mimicry by making people less able to get closer to other people's emotions.

Egocentrism is akin to vanity, the sin of pride, or the *hubris* that Greek gods observed in arrogant men carried away by the intoxication of power.

Omnipotence and Narcissism

It would therefore be naïve to deny the existence of this self-centered bias, but some leaders manage to push it very far when, once in control, they think they can act according to their own desires without following social rules or conventions. A certain amount of narcissism can be accepted for those who exercise responsibilities, because they did not get where they are today by being modest. But self-promotion and self-confidence lead some executives to accumulate power and seek to expand their sphere of dominance regardless of the financial consequences when they engage in grandiose merger and acquisition projects that run a high risk of failure.

Difficulty of Perseverance with the Bias of Immediate Reward

What undermines the credibility of leaders is not only their tendency to put themselves first, but above all their lack of persistence and long-term commitment.

In the previous chapter we saw how long-term thinking is constrained by the bias of immediate reward. Long-term strategic thinking is frequently constrained by the power of shareholders and the need to show convincing and immediate results. This often leads to frequent changes of direction that bring instability and rapid turnover at the head of organizations. Yet, a change of direction can only succeed if it can be maintained over a fairly long period of time. But this is not what is generally observed as reorganization projects rarely last longer than the tenure of the leader who initiated them, and it is not certain that a transformation program that is beginning to bear fruit will continue with the arrival of a new boss. The newly appointed leader will seek to leave his or her mark with visible actions that will distinguish them from previous programs.

Given the lack of long-term perspective, charismatic leaders are often called in, with the hope that their charisma, that is their understanding of the situation, will fill the gaps and mobilize people. Heroic stories about their successes or failures, however, remain very simplistic when considering the complexity of the situation they must face. If the results are good, leaders are acclaimed and they must be retained, but otherwise, charisma becomes incompetence, and they are dismissed and replaced. Most of the time, 'heroic' leaders succeed in the short term by resorting to drastic decisions, disrupting organizational charts, or restructuring in ways that weaken the quality of the interpersonal relationships which are so important for good collaboration and the achievement of long-term, convincing results.

In conclusion, long-term engagement and dedication of leaders is a very powerful motivator to mobilize followers, but to stay the course and avoid the risk of seeing the creative model falling back on the fixed model, leaders must understand and manage their biases while engaging over the long term and leading by example.

Sharing the Vision to Mobilize Followers

As strategic vision and priorities are only useful if they are shared to guide action, leaders must find the right way to communicate them and explain their rationale and consequences. This sharing will be communicated more effectively if the team members can participate in the reflection from the start. To be sure that the message is heard and understood, information is not only passively transmitted, but also actively shared to allow everyone to question the direction followed and anticipate what is to be expected, even if the future is not rosy.

Transparency and Repetition

Communication should reinforce the core priorities, the few things that everyone should grasp to conduct their mission. To grab attention the same message is repeated over and over in different forms. Say it a few times, people do not even notice. Repeat it fifteen to twenty times and people start getting it and wiring it in their brain. Moreover, leaders should trust employees with all sorts of vital information and go against the temptation of hoarding information as a way of gaining power.

Telling a Convincing Story

A good way to communicate and share the strategy is to translate it into a simple narrative that can be easily associated with people's interests and the concerns already present in their heads. This association will help them to better connect with the situation and understand it. We have an insatiable need to bundle facts and ideas into a coherent and familiar story. And once this story is settled on, we seek to confirm it. The story will explain the reasons for the choices and pinpoint the problems to be solved. It will clarify the starting and the end point of the journey, how things will unfold and improve, and how everyone can contribute and make an impact. This is a way to activate internal motivations, as we know that people look for meaning and need to feel part of something exciting, with sufficient autonomy to master what they do. They also need to belong to a united community that shares the same culture and the same values and that includes and respects them.

It is not always easy to give meaning to one's work, as many tasks seem to bring little interest and added value. Many unnecessary activities occur which should be phased out, like repair or inspection activities when quality is inadequate, or repetitive tasks that could be automated. To regain a sense of meaning in this type of work, the focus should be on the analysis and elimination of this type of activities.

The Gap Between Vision and Reality

The problem with well-designed strategies is that they often remain at the exhortation level far from the concrete stage of implementation. Speeches and hype create a gap between talk and reality. The value statements emblazoned above the entrance hall or in annual reports often turn out to be rather elusive and abstract formulas. They are supposed to provide meaning and reassurance, but these values in their succinct formulation hardly reflect actual practice and how they impact the working relationships of employees in the context where they are placed.

For many years a very large American company proudly displayed the following values in the lobby of its headquarters: "Integrity. Communication. Respect. Excellence". This was the motto of Enron, a company which collapsed amid a shocking scandal, mired in fraud, and institutional corruption.

Similarly, performance and results are often displayed in quantified and financial terms, based on major strategic visions which hover above the concrete reality of the situation experienced in the field. It is not by maintaining an abstract communication approach, at a remove from the internal workings of the organization, and by focusing on immediate results, that leaders will convince people to engage with enthusiasm.

Mass Empowerment. Everyone as a Potential Leader

Empowering

Showing one's commitment and sharing the strategy and priorities are not enough to ensure the support of followers if leaders are not ready to share power with the various actors and reduce their obsession with control and immediate results. The objective is to move from a leader–follower to a leader–leader mode of action by empowering collaborators and making them responsible leaders, from senior managers at the highest level to operatives on the frontline. Empowering talented people consists of giving them a sense of freedom to achieve something meaningful. Responsible people thrive on freedom and decide how to organize and improve team work.

Letting People Take Ownership of Their Work Processes

One way to avoid returning to the comfortable mode of command-and-control is to have the *authority* of responsible actors enshrined in the structure of the organization to make them independent from the goodwill of their superiors. They take ownership of their work processes and can seize opportunities for innovation or improvement, as they happen. They have the appropriate means to measure and control what they do, and they no longer need to pass the many performance indicators of their activity to their superiors. Information relevant to control and improvement is close to the action, and they have the freedom to try new things and potentially learn from their errors.

Better still, they can use their autonomy to broaden their scope of responsibility and cooperate with others, exploring reciprocity and developing new competencies. As Laslo Bock,[3] former head of people operations at Google, explains "For over sixty-five years, 3M has offered its employees 15 percent of their time to explore… Our version is 20 percent time, it means that engineers have 20 percent of their week to focus on projects that interest them, outside of their day jobs but presumably still related to Google's work (at Google that still covers a lot of territory)".

Taking Power Away from Managers

At Google there is deep skepticism about management, and they deliberately take power and authority away from managers by letting them make *few decisions unilaterally* and by eliminating status symbols. There are only four meaningful, visible hierarchical levels: individual contributor, manager, director, and vice president. The most senior executives receive only the same benefits, perquisites, and resources as the newest hires, and there are no executive dining rooms, parking spots, or pensions.

[3] Bock (2015).

The Leader as Coach and Integrator

But in fact, managers do many good things. Laslo Bock pinpoints "Managers serve the team…They do not focus on punishments or rewards but on clearing roadblocks and inspiring their team". Leaders practice new roles as knowledge transmitter, coach, trainer, or advisor, helping people develop teamwork and cooperation across units. To assist employees in fully owning their working processes, they organize exams, audits, or certification procedures to make sure that things are running smoothly and with the right level of competence.

Are Leaders Ready to Let Go of Control?

Empowerment is certainly a powerful motivator, but it is the hard thing to do and is difficult to manipulate. It is much less comfortable to explain to followers why they should do something rather than tell them to do it. Moreover, when they accept delegating power, leaders must give back some of the hard-earned power they have accrued in reaching their current position. Are they ready to take a step back and let go of control?

Are They Ready to Trust Their People?

Most managers are apprehensive of letting go of control, of having to debate in case of disagreement, of taking risks for their career if things go wrong. So, they micromanage to get reassurance by controlling the actions of their staff. This shows a lack of trust in the competence of their people, with the fear that something may go wrong.

After all, leaders are supposed to know, why should they accept delegating the decision? According to the bias of egocentrism and overconfidence, they know better than others and they trust more their own decisions. The easy path is the command-oriented style, which is more natural and requires less effort as it is in line with the hierarchical bias.

The Strength of the Hierarchical Bias

We cannot fool ourselves; hierarchical bias is always present even if it is latent. This bias remains pervasive in most institutions and leads many individuals to fight for status and gain power to maintain it. Social intelligence will be needed to identify the sources of power, understand the actions of others, intimidate a rival, or muster enough control of one's impulses to ignore most provocations. The political game requires the formation of alliances and coalitions that will make it possible to rise above others and gain prestige and status. It is not possible to break free from it, nor to avoid political games, even though these dynamics produce stress and reduce job satisfaction. As this climate does not encourage employees to take risks and bet on the long term, they will simply remain focused on results that are visible and easy to achieve.

What is going on in the brain? Naturally, the prefrontal cortex, the PFC, is at work, as evidenced by the fact that lesions to this region make patients unable to recognize relationships of dominance or submission. The mentalization system activated in the PFC, and the upper temporal gyrus, also shows how we read the minds of others to acknowledge their beliefs and understand how they reason. In situations such as negotiations, this ability to understand the other's point of view provides a distinct advantage, even greater than empathy, because it allows us to anticipate what is going on in others' minds.

Hierarchical bias

Hierarchical organization is inherent to the social nature of species, as we saw in Chap. 1 with chimpanzees where alpha males dominate their community. Like many other species, we have an implicit sense of hierarchy, which is based on thousands of years of genetic evolution and social conditioning when the need to cooperate in bands or groups emerged. Even in the absence of a formal organization represented by titles and roles, members of a group willingly organize roles in a hierarchical way.

As we began to accumulate resources and possessions with the development of animal domestication and agriculture, we invented some modes of subordination that far exceeded those present in primates. Civilizations developed stratified and rigid social hierarchies that distributed power and prestige in a highly unequal way. For example, the Babylonian social order was based on three well-defined classes: free men, commoners, and slaves.

Thus, hierarchy signals position and status and engenders a classification system that formalizes unequal access to limited resources, be they monetary, physical, or status related. We continually emit waves of messages to recalculate our status in relation to others, and all the signals we receive, whether positive or negative, accumulate to define our position in the group.

Domination and Social Status

The lure of social situations of domination and power has continued to attract and fascinate for the prestige and pleasure they provide, but also for the considerable benefits they procure. So, successful leaders will make great efforts to protect, strengthen, or expand their situation and seek out potential allies to maintain it.

This obsession with hierarchy and social comparison has been embedded in our neurons for far too long for us to hope that it will fade anytime soon. Research by Caroline Zinc[4] shows that hierarchy is wired into the human brain just as it is in primates. Our primate instinct is accustomed to measuring existence according to prestige, hierarchy, and social recognition. We are very concerned with knowing where we are in the hierarchy, and we are very expert at spotting differences in status at a glance. It only takes 40 ms to spot a dominant face with a direct and assertive look or, conversely, a face offering an elusive gaze signaling submission. Watercooler talk feeds on who has risen in the ranks, been demoted, or made the biggest blunder.

Complex neural maps in our brain represent our own status, and the hierarchical order of the people around us, in multiple dimensions, such as age, income, physical strength, or sense of humor. Physical size continues to unconsciously influence our perception of domination even in the most advanced democracies. In the United States, just under 15% of men are taller than 6′ (183 cm), but this percentage rises to almost 60% for Fortune 500 CEOs.

Conversely, hierarchy can satisfy the need of some people to be cared for and guided. Hierarchy establishes a stable, secure structure, which provides certainties on appropriate values and attitudes. The world is clear and organized, and people accept submission to this order. Chapter's experiments have shown what people are ready to accept in the name of obedience and conformity.

Dopamine and Testosterone

Rising in the hierarchy feels good and releases a spurt of dopamine in the rewarder. Moreover, reaching a position of power also leads to an increase in testosterone which pushes leaders to do their best to maintain their status, while reinforcing their self-confidence. Does this testosterone rise lead to an increase in libido, as witnessed sometimes with the multiple affairs and sexual scandals of certain bosses, presidents, or stars? That remains to be proven. Nevertheless, this burst of testosterone reduces or inhibits the synthesis of oxytocin, the neurotransmitter associated with the ability to trust others. The result is that bosses become more selfish, less empathetic, and interested in others, especially those below, who are looked down upon.

So, It is not Easy to Take the Hard Path and Trust People

In summary, a leader–leader style requires determined commitment over time, sharing of the strategic vision and the desire to redistribute power while mitigating the

[4] Zinc et al. (2008).

hierarchical bias. But this is not enough as followers must be trusted and competent and placed in the proper context and culture.

As employees are given a greater sense of control and authority, they must become more accountable. So, accountability requires selecting and recruiting great people that have the proper competences and can be trusted, two aspects that will be considered in the next chapter dealing with context and culture.

References

Berlin, I. (1953). *The hedgehog and the fox: An essay on Tolstoy's view of history*. Weidenfeld and Nicolson.

Bock, L. (2015). *Work rules! insights from inside google that will transform how you live and lead.* John Murray.

Zinc, C. F., Tong, Y., Chen, Q., Basset, D. S., Stein, J. L., & Meyer-Lindenberg, A. (2008). Know your place: Neural processing of stable and unstable social hierarchy in humans. *Neuron, 58*, 273–283.

Chapter 9
Establishing the Right Context and Culture

Hiring the Right People

Importance of Cultural Context in Mobilizing Followers

However brilliant and charismatic leaders are in defining a strategic direction, and exercising an effective leading style, they must get the right people on board, set up the proper context, and organize a culture of relational engagement that will support their approach. This is well illustrated by Tony Hsieh, the chief executive of Zappos.com who explained in an interview "I think of myself less as a leader, and more of being almost an architect of an environment that enables employees to come up with their own ideas, and where employees can grow the culture and evolve it over time".[1]

Hiring the Right People and Developing Competencies

To empower people, you need to trust them. You must check that they have the right competence for the decisions they make and that they are readily engaged in what they do. *Competence* and *engagement* are two key factors that must go along with power delegation. Therefore, leaders must have an eye on who is hired and who is promoted and check that these people are competent and motivated to manage and improve their own processes.

So, it becomes essential for the organization to control the talent it employs and get the right people on the bus and the wrong people off the bus. According to Laslo Bock[2] former head of people operations at Google, "Hiring is the most important

[1] Bryant (2010).

[2] Bock (2015).

© The Author(s), under exclusive license to Springer Nature Singapore Pte Ltd. 2023

J. Teboul and P. Damier, *Neuroleadership*,
https://doi.org/10.1007/978-981-99-5122-2_9

people function you have, and most of us aren't as good at it as we think. Refocusing your resources on hiring better will have a higher return than almost any training program you can develop". And he adds "We believe that hiring should be peer-based, not hierarchical, with decisions made by committees, and it should be focused on bringing the best possible people into the company, even if their experience might not match one of the open roles".

Emotional and Social Competence

When selecting and recruiting new employees, retaining the best talent certainly requires being attentive to their mastery of the relevant competences needed for their job. But it is also essential to check that they can manage themselves and are able to collaborate with others. Do they have the emotional intelligence that will guide them to develop a culture of collaboration and learning? Do they have the social fiber that will allow them to organize the necessary cross-disciplinary cooperation?

Diversifying the Group

When hiring, it is also important to be aware of an inherent bias that appears in the functioning of any group, *the bias of similarity and social conformity.* We have a tendency to trust our loved ones, those who look like us, those we are familiar with, or belong to our tribe. We are suspicious of others, the strangers who pose a threat because they are different. This bias can lead managers, when they hire new people to prioritize candidates of similar age, personality, or educational background to their own, rather than candidates who are different and may challenge them.

This bias of social conformity is a serious drawback when it comes to innovation, as we know that it is essential to have teams as varied as possible to support and enrich creation. Hiring "super-experts" from the same field of expertise to make a powerful difference in numbers, most often generates less questioning and creativity and produces inferior results to an open recruitment of people that offer a more diversified skillset.

Engagement

Engagement is dependent on motivation, and we have seen, in Chap. 5, the importance of playing with extrinsic and intrinsic motivators. Salaries, bonuses, as well as perks and amenities have a direct effect but the search for meaning, the sense of freedom, of being included, informed, and recognized can have a more lasting effect in mobilizing people.

Setting up the Proper Context

After sharing their vision and then trusting and empowering their team, leaders must establish the right context to reduce hierarchical sluggishness and organizational complexity. This deliberate approach is built step by step throughout the organization and must be deeply embedded in the organizational structure. For example, freedom of action for staff is inscribed into the system, independently of the goodwill of their superiors.

Lighter Structure and Redistribution of Power

To give operators more visibility and control over their own processes, the main challenge is to flatten hierarchies, limit the proliferation of rules and procedures, and reduce the complexity of cross-functional interactions. There are far too many coordinators, functional managers, or intermediaries who prevent better contact with field operations and transversal cooperation across units. Similarly, the accumulation of rules and procedures does not give employees sufficient autonomy to take advantage of opportunities for improvement and innovation. Delegating power reduces the need for reporting abstract and financial performance indicators upwards.

In short, leaders should strive to simplify and reorganize the context by having the problems addressed where and when they arise and avoiding higher level reporting and approval. This leads to more concrete and specific action with less bureaucracy, less paperwork, fewer meetings, and more rapid decisions. The more freedom employees get, the more they can draw on personal motivators to engage and cooperate.

What Are the Obstacles?

Setting up such a favorable context is not easy because the management systems in place pull against freedom. Even when managers are aware of the value of delegating power to make employees accountable for their work, they nonetheless find a way to achieve direct centralized control by using information technologies. These technologies shorten the distance from the top to the base, and many intermediate layers are replaced by sophisticated computer programs that enable the instant recording of all kinds of operational performance indicators, or personnel situations. They show in real time, on managers' screens, the development of resources and the results achieved, which speedily convinces them that they know enough to take decisions from their command post. The dashboards on the screens replace practical experience in the field, and distance leaders from the role they should play as coaches and integrators. Once again, employees collaborate under the watchful eye of multiple reporting systems and focus their attention more on metrics related to predetermined objectives than on being concerned with what is best for the company.

Practicing a Culture of Relational Engagement

Relational Engagement

Developing *relational engagement* is a way to mobilize the whole organization when people get to know and educate each other for the success of all. The drive to make steady, consistent progress toward a common goal gives everyone a feeling of belonging to an active and united community. Transformation does not happen in one fell swoop, but in a relentless and disciplined move to build momentum through practice and experimentation.

Empowerment needs trust. Trust is established through a reciprocal engagement as people know each other better and are more predictable. To build a good working relationship with a coworker, an employee, or a supervisor, it is therefore essential to consider them as a real person, with whom a *personal relationship* can be developed around a shared experience. It is not about being nice to each other, or even becoming friends. Fully fledged relationships require mutual adjustment efforts by avoiding indifference, manipulation, or concealment. If asked for help, you respond immediately as the commitment to learning and collaborating takes precedence over the transactional, give-and-take mode where people remain focused on their short-term personal interests.

Geographical distance, specialization, and power dynamics are all barriers that hinder relational engagement and suck energy from the system. They lead to high transactional costs and greater management complexity. Computer scientists, salespeople, or accountants think differently and have different values, different timelines, and jargon. As they tend to remain *fixated* on their roles and modes of operation, coordinators and multiple intermediate structures are needed to compensate for the relationship difficulties between different cultures. The goal is therefore to try to exit this dry and formal transactional mode and regain some agility, thanks to real and *personalized relationships* that allow a spirit of trust, rising above differences.

The culture of an organization stems from regular shared behaviors in daily exchanges. People become more predictable and familiar as they interact through regular meetings or experience-sharing sessions. These behaviors are transformed into norms and habits through repetition. The more these habits and norms are internalized and etched into the neural circuits of people's brains, the more informal and natural the culture becomes. Multiple formal regulations and intrusive supervision become less necessary.

Employees steeped in this informal culture gain autonomy. Their behavior becomes predictable, and they can anticipate the behaviors of other members of the organization. As the brain is an *anticipating machine*, knowing what to expect from others is reassuring and increases cooperation. *Comfort in predictability* is another definition of trust. This mode of relationship allows groups to be formed and reformed easily, depending on needs or circumstances, because each actor is attentive to others and understands how to adjust.

To avoid rigid roles and transactional relationships, managers may be rotated to give them the experience of walking in others' shoes. At Google it is expected that various people will step into a leadership role over the life of a project, as different skills are needed at different times.

Performance Springs from Emotional and Social Intelligence

Researchers from Carnegie-Mellon and MIT found that the best performing groups were not those with the brightest individuals or those with the highest overall level of intelligence, but those with three main characteristics: a high degree of *social sensitivity and empathy*, an *equal speaking time* (no one dominates or stays silent), and a *higher proportion of women* (given the greater social orientation in women as opposed to men). We are not surprised to see the importance of *emotional and social intelligence* in an environment which is all too often overly rational and transactional. Human bonds facilitate and accelerate discussion between players, allowing them to combine their talents and learn faster. It is less about developing knowledge than building strong teamwork and networking capabilities. When people know each other well, they can go beyond prescribed roles, support each other, and know who to ask for help. We need mortar, not just bricks.

At the consulting firm McKinsey, consultants can ask their colleagues for help with the assurance of receiving a quick positive response. And of course, this willingness to respond to requests from other members is an important element of the appraisal process. The collaborative culture cements internal cohesion.

Strengthening Trust

A culture of relational engagement requires a high level of openness and trust, but this collaborative mode could rapidly revert to a transactional mode if some players no longer comply with the rules of the game. It follows that selfish and transactional behaviors must be regulated by internal feedback mechanisms and eventually sanctioned. A culture of sharing and cooperation is not driven by angelic altruism, but by a calculation of reciprocity, a balance between individual interests and those of the group. To reinforce the collective process, fairness is important. Is everyone doing their part or are there a few freeloaders taking advantage of the situation? People are more ready to engage when they feel that they are equitably consulted and involved in decisions, and deviants are penalized.

In conclusion, it takes a long haul to establish a fluid mode of interaction and to escape the culture of immediate results, transactional relationships, and struggle for power. Trust is the glue that maintains social relationships and facilitates exchange. But this fluid collaboration is not enough, a *learning culture* is also needed to enable continuous improvement and innovation. But before considering the learning culture, it is important to review what limits relational engagement.

What Limits Relational Engagement?

Relational Engagement Goes Against the Bias of Least Effort

For the creative leadership model to function, people of different professions, approaches, and positions must accept collaborating within an informal culture of relational engagement. But relational engagement is not easy to sustain as it requires everyone to pay the price in time and effort to adjust to the other person. It's much easier to let go according to the *bias of least effort* and thus let relationships fall back on a partial and transactional mode where exchanges take place according to the roles of the task at hand and remain distant and policed. The fluidity of exchange is lost as the culture becomes more formal.

When we observe the way meetings go, especially at a high level of responsibility, we see that everyone stands on their territory, defends their own interests, and the exchanges come to resemble a game of ping-pong.

Relational Engagement Goes Against the Bias of Social Conformity

We have already noted the formation of vertical silos in the pyramidal structure as people come to work together according to their local objectives or their specialization. They defend the interests of their silo against other silos. This "us versus them" mentality results from the *bias of social conformity* and poses the nagging problem of transversal coordination between units to respond to the final demand.

The bias of similarity and social conformity

Genetic and cultural evolution has favored socialization within the protective framework of the family or the tribal unit. These culturally homogeneous groups had a size that did not exceed 150 individuals[3] to favor close collaboration and allow all members to know each other in face-to-face exchanges. The adaptive benefit of trusting community members was essential for comfort and protection. Familiarity and belonging to the same mental world cost less effort in relating to others and provide a feeling of security. As a result, we seek to associate with those who look like us, whether in profession, cultural habits, status, or language. When we are among peers, we share the same views, we forge strong bonds, and our identity is well established. We are thus conditioned to seek social acceptance, and we are willing to make a lot of efforts to be included in the group. A sense of belonging to a group is as basic a need as drinking or eating. Conversely, exclusion or ostracism can engender the same pain as thirst or hunger.

Like other species we automatically distinguish between 'we', our loved ones, those from our side and 'them', the strangers, those from the other side. And we consider that we are superior and more trustworthy than the members of the other group who are viewed with suspicion or even rejected.

Basic sensory stimuli that differentiate one group from another are often minimal and rapid, as they result from emotional processes. Only 50 ms of subliminal exposure to an unfamiliar face, or the sight of someone from another racial group, is sufficient to activate the amygdala. As the exposure is extended long enough to allow conscious detection of the face (about 500 ms or more), the initial activation of the amygdala is followed by the activation of the reasoner in the prefrontal cortex that moderates and calms the amygdala. The prefrontal cortex is again called upon to regulate embarrassing or threatening emotions.

The bias of social conformity leads to *groupthink* as individuals tend to conform to the opinion of the group, to the majority, rather than risking an open conflict. In the name of compliance, they are led to accept absurdly incorrect propositions.

In the 1950s, the social psychologist Solomon Asch did a series of experiments that showed how an individual decides when faced with a group whose opinion differs unanimously. The experimenter asked small groups to compare the length of a line with that of other lines projected onto a screen. Each participant had to respond in turn, aloud. The subject was the last to answer, and the

[3] The anthropologist Robin Dunbar has shown that there is a correlation between the size of the social group in primates and the size of the neocortex of the species in question. The necessity of managing increasingly complex social interactions led to the increase in the size of the brain. For instance, chimpanzees live in communities of between 50 and 55 individuals. Dunbar has estimated, by extrapolation, that the typical size of a human group where everyone knows one another personally is 150 individuals. This size, which has remained stable since prehistoric times, can be found today in personal mailing lists for greetings cards or in the number of friends on social networks.

other participants, in cahoots with the experimenter, all confidently announced the same wrong choice. The result of these experiments showed that the pressure exerted by the group's unanimous thinking was sufficient for most of the subjects to conform to the collective opinion, even if the group was composed of complete strangers.

As a result, the bias of social conformity hinders good collaboration and coordination between groups by creating an Us versus Them mentality. The other downside is that managers tend to recruit people who have the same background or training, people who are less likely to oppose or criticize the existing social norms. Conformity and thinking homogeneity also lead to poorer decision making as the group is not able to benefit from a diversity of experiences and perspectives of its members, which is an important source of creativity. The dominant and well-established ideas are self-reinforcing and converge toward a single type of thinking which may be far from optimal, especially when the environment is changing rapidly.

Practicing a Learning Culture

Experimentation

Learning proceeds by trial and error and requires some tolerance for mistakes. It is the dopamine circuit that pushes us forward to learn from our errors. The surge of dopamine we receive for identifying an error stimulates and reinforces our exploratory behavior. Learning starts from what doesn't go *as expected*, whether it is a defect on the assembly line, a new drug that fails during the clinical trial phase, or a meeting that does not reach a conclusion. So, experimentation makes us take the risk of being wrong. But, if we are penalized for failure and not rewarded for taking risks, we opt for safety and let the bias of fear and insecurity take over. And in the event of a misstep, we do our best to hide it.

Therefore, creative leadership requires a climate of *psychological safety* which allows people to tolerate failure in order to learn faster and to carry out analysis of errors without blaming anyone for it. The rollout of multiple prototypes is a good way to "fail often to succeed sooner".

Learning from Experience[4]

What is the secret to your success?

- Making the right decisions.

  What helps you make the right decisions?

- Experience.

  What provides you with experience?

- Poor decisions.

Self-affirmation and the Courage to Speak Up

To cultivate a climate of psychological safety in daily operations and see initiative flourish, it is also essential to allow people to assert themselves and have the courage to speak up, flag up errors, and potentially contradict their superior without being penalized. It is important to fight against the force field created by the hierarchical bias and be able to voice questions about how things are run without worrying how this will make you look. This is an essential source of progress for the group.

To develop a learning climate, leaders should engage in dialog and debate, confront brutal facts, and hear the truth from their own people. Will they be humble enough to explore their mistakes and personal weaknesses?

Learning Everywhere, All the Time

The need to develop and increase the level of competence leads to the simple reality that everyone must learn every day. Learning becomes active when related to daily work. This basic recognition unifies all actions and is the engine of continuous improvement that makes the most of the hundreds of opportunities that arise every day. Presentations, lectures, or seminars are useful but remain a passive mode of learning.

The basic mechanism that ensures competence and its maintenance is to learn everywhere, all the time. But if learning is all that happens, when are people supposed to work? The answer is obvious, learning is done by doing. Learning helps grow every day, at work and in life. Managers and supervisors are there mainly to help, guide, and educate, as we saw in the Toyota case.

---

[4] Stern (2019).

Embedding Responsibility Within the System

Let's take the classic case of a nurse who doesn't dare speak up when her boss, the surgeon, misses a step in a care protocol, because she is afraid of being rebuffed. Training the nurse to self-assert and speak up would be a waste of time if the context is not supportive. It is not the behavior that needs to be changed, but the context itself and the operating process. By making the nurse responsible for monitoring the operation protocols and attaching this responsibility to her position, she now has the authority to intervene with more peace of mind. This responsibility is embedded into the system and no longer depends directly on her superior.

What Limits the Learning Culture?

It is Difficult to Unlearn

The experience we gain when we learn provides us with some functional comfort as we often act and decide according to well tested and automated behaviors that become habits, and that are wired in neural circuits in the brain. So, change is stressful and effortful because we must unlearn first, to avoid taking the circuits already established in the brain. Then the creation of new networks requires a lot of attentional energy as we proceed slowly with frequent trials and errors, before gradually relying on automatic behaviors and habits. In fact, it is more difficult to unlearn than to learn (see Annex 7). The result is that change is not welcomed as we are confronted by the bias of least effort that makes us revert back to the old ways and the return to status quo.

People Are not Very Flexible

In a rapidly changing world, organizations need flexible people who can evolve and quickly turn the page, people willing to abandon what they already know and adopt new frameworks and specific demands. They must learn to forget, learn to unlearn, learn to change the way they frame the world. But most people are not that flexible. They need benchmarks and guidelines. They need to build on what they have already achieved. Unlearning and taking a new direction comes at a cost and goes against our predisposition to confirm our past beliefs and opinions. A problem common to innovators is that they often try to protect their personal achievements from the threat of new ideas. They remain trapped in their past successes, the contracts won, or what pleased their best customers.

The Insecurity Bias Hinders the Learning Culture

The insecurity bias is always present during learning, which occurs mainly by trial and error. It does not encourage initiative as everyone is afraid of making mistakes or being at fault. Fear takes precedence over confidence: the fear circuit activates the amygdala and limits the normal functioning of our prefrontal cortex. Since there is an association between error and punishment, we hide our failures and avoid venturing too far into unknown territory when learning from experience.

Insecurity also prevents employees from speaking up freely. They remain silent instead of asking for explanations or giving their opinion. They are not prepared to willingly tell their boss the difficulties they faced when trying to apply what they were asked to do. They do not want to be the bearers of bad news and know that questioning established beliefs or interests dear to their superiors can come at their expense. Better not to see and not to know.

Orchestrating the Whole

Rewards. Victory is Collective

Measuring results and recognizing the contributions of everyone are the counterpart of autonomy. Measurement of performance must go beyond formal and abstract systems that assess short-term results, as a well-run effort does not necessarily rapidly produce convincing results. This is accentuated by the fact individual indicators are easier to measure than the assessment of collective effort and teamwork. Moreover, when it comes to giving feedback, the social dimension is often overlooked. Reward and promotion systems should consider collective action and effective cooperation, as well as individual performance. Since it is hard to measure cooperation, its recognition depends on the observation and judgment of operational managers who know what is going on in the field. They can gather concrete data and give direct feedback instead of waiting too long to make generalized judgments, far-removed from daily experience. Moreover, some form of discipline and progressive sanctions should be applied when the social pressure of the group is not sufficient to enforce accepted rules.

Over and above compensation and intrinsic motivators that we have mentioned, celebration and public recognition of accomplishments can add emotion and joy. That is a good way to make things memorable.

The Big Picture

The creative leadership model and its cultural transformation concern the organization as a whole and must operate in a systematic and deliberate way. If not, proceeding in a piecemeal, fragmentary way or resorting to mere exhortation leads to wishful thinking.

Coordinating the Whole Process

The proliferation of local initiatives should be harmonized and aligned to goals by a coordination cell that will also support educational efforts. This facilitation structure will not be responsible for carrying out the work, which is the realm of operational managers in the field. It will accelerate the movement by increasing the numbers of experiments that will transform behaviors, strengthen the culture, and develop competences, through regular practice. This coordination cell will disseminate suitable tools and training programs to develop a culture of experimentation and collaboration, as illustrated by the Toyota example in Chap. 6.

What Limits Overall Evolution and Orchestration?

Rapid Evolution of Competencies

The creative model of leadership should be able to confront the current economic environment where globalization, automation, and new technologies drive change and convey new threats for employees. Globalization brings in competition from overqualified, lower-paid workers. New technologies force employees' competences to evolve rapidly to keep up to date. Acquired experience no longer provide a guaranteed position as people must be trained and retrained regularly. Age also comes into play when, for example, it comes to choosing between hiring a 25-year-old new employee or retraining a 50-year-old worker.

New candidates must accept the challenge of renewing themselve, with less time to acquire professional competence and develop relational engagement. Their acquired experience may be less important than their ability to reprogram themselves and adapt to others. Those who are successful are more autonomous and have enough potential to reinvent themselves. But how to manage and include those who are afraid of becoming useless or risking being made redundant? Our neural circuits are harder and harder to reprogram as we grow older, and the insecurity bias plays a major role here.

Furthermore, a certain level of diversity within teams is required to question the status quo and innovate. This is not obvious as the bias of social conformity leads us to hire people who have a similar education or outlook on life, rather than candidates who have different views and are prepared to challenge us.

Reward Remains Individual and Blind to Long-Term Effort

Performance assessment systems are often formal and typically consist of readily available individual metrics which do not sufficiently consider the level of team collaboration and mutual effort. Cooperation is difficult to measure and should be assessed by direct, field-based observation of people's daily contribution. But instead of this direct feedback, those in charge often wait until the end of the year to make a general cursory judgment as part of the notorious performance review.

A Disjointed, Piecemeal Global Approach

Certainly, there is a will, at the highest level, to mobilize the whole organization toward the chosen directions but this determination quickly erodes across hierarchical layers when discipline and motivation falter. Management often resorts to exhortation with big announcements and ambitious action plans that go with training and improvement programs that run through the whole organization. But most of the time, these approaches remain disjointed and piecemeal, and they focus on tightening the bolts and obtaining immediate and visible results. Moreover, leaders are not keen to stay in place longer than necessary, and you can expect that any new boss will soon steer the organization in another direction. In the absence of a sufficiently powerful coordination structure to orchestrate the whole over the long term, the organization will default to the fixed model of leadership.

References

Bock, L. (2015). *Work rules! insights from inside google that will transform how you live and lead.* John Murray.
Bryant, A. (2010, January 9). On a scale of 1 to 10, how weird are you? *New York Times.*
Stern, S. (2019). *How to: Be a better leader*. Macmillan Publishers.

Chapter 10
Accepting Our Brain as It Is

Back to the Overall Picture

A Brain with Some Pitfalls

Human beings are complicated, but they can be trusted when they are motivated and able to deal with the limitations and proclivities of their brain. We must take the rough with the smooth, forget Rousseau's mythical 'good savage' [1] and remain aware that the very functioning of our brain, which can achieve extraordinary feats, can at times present some serious pitfalls. It is very unfortunate that most leaders today remain unaware of the systematic inclinations that profoundly and unwittingly influence them. Our education system, which is mainly concerned with making us rational and logical beings, remains almost completely silent on the topic. So, it is about time to take stock of the main aspects of the functioning of the brain before reexamining the seven capital biases that we have touched on in previous chapters.

Memory Notepad at the Conscious Level

The 7/3 Window

Thinking involves juggling with mental items in the working memory, a notepad which is able to scan our long-term memories (Fig. 1). This narrow window of consciousness is only able to consider, on average, up to seven pieces of information

[1] Rousseau starts from the idea that man is good, nonviolent, and self-sufficient by nature and that civilization perverts and distorts him.

© The Author(s), under exclusive license to Springer Nature Singapore Pte Ltd. 2023

J. Teboul and P. Damier, *Neuroleadership*,
https://doi.org/10.1007/978-981-99-5122-2_10

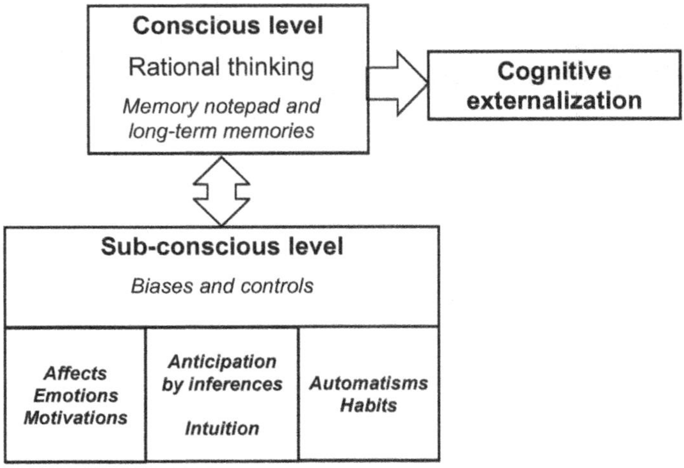

Fig. 1 Three levels of information processing

at a time and to readily deliberate with only three.[2] Thus the 7/3 rule: *Consider up to seven elements and work and deliberate with three.*

Too often, to be more convincing, we want to be exhaustive and deal with long lists of arguments. This is a mistake. Just keep it simple by focusing on the few points that matter most. What are the key challenges, priorities, main lines of action? Where are the bottlenecks? The working memory will not be overwhelmed if we stick to this limit of about three elements, knowing that it is always possible to cluster or regroup information (it is as easy to remember three words as three letters).

Avoiding Cognitive Overload of the Prefrontal Cortex, More is Less

When the prefrontal cortex is under heavy strain, for example, in situations of intense cognitive work, intense negotiations, or difficult social interactions, the activity of this area decreases and working memory is saturated. Performance declines rapidly and gives rise to stupid mistakes or poor choices. Empathy is reduced, and people tend to cheat. Emotions can take over, leading to anxiety and frustration (Fig. 1).

[2] The number three is often used to articulate memorable representations: the Christian doctrine of the trinity, the three French values of "Liberty, Equality, Fraternity", the three colors of the flag, the three leaders of a roman triumvirate, Neptune's trident, the Hegelian dialectic of thesis, antithesis, synthesis, and even, the three Musketeers, the triple time of the waltz, or the three blows hit on the floor before the beginning of a theatrical performance….

Affects and Emotions

Emotional Reasoning

It is important to remain aware that our feelings can deceive us when we base our opinions and decisions on our emotional perception. What we feel colors what we think. *Emotional reasoning* not supported by additional evidence could influence our choices.

Emotional Intelligence and Self-control

The region of the brain which is regularly involved in regulating and controlling our emotional states and impulses is the OFC, the orbitofrontal cortex[3] (see Annex 5). But the braking circuits quickly reach their limits, like a fatigued muscle. We must therefore be sensitive to emotional overload and stress which might disrupt or block our emotional regulation. As the amygdala takes precedence over the frontal lobes it can overwhelm the cortex in a flash, thrusting us in a state of mental blindness. We lose control. One way to reinforce this control is to develop automatisms and habits with regular training (this is in fact what politeness and rules of conduct are about). Sometimes it is necessary to resort to imposed restraints or threats that leave no choice but to submit.

Self-control and Reframing

In fact, self-control seems to work more by reframing perception in a productive way than by exercising direct control. Thus, the key to managing emotions might be to step back and reappraise the situation to calm the emotional response before it becomes unmanageable. This form of top-down control can be done by labeling the situation or reappraising it in cognitive terms to diminish its emotional intensity. *Cognitive reappraisal* is based on the idea that what makes us emotional is not the situation itself but the way *we think about it*, the way we represent and *frame* it. This is what happens to optimistic people who see life through slightly rose-tinted spectacles.

Inferences and Intuition

Intuition can be trusted when we have long practice with similar situations in a stable and regular environment. But when the context is turbulent, and the feedback ambiguous or delayed, accumulated expertise can be misleading and real expertise

---

[3] The orbitofrontal cortex is situated in the lower third of the prefrontal cortex, above the orbit.

is very difficult to develop. Since we have no access to the processes underlying intuition, mistakes are difficult to correct, and we can easily fall into the *trap of overconfidence*. The risk is therefore to see consultants and experts giving in to confirmation bias by seeking to impose their views. Contradictory information is ignored, and post-rationalization is used to reduce dissonance. So, as a novice in a new field, you are better off using rational reasoning.

Habits

It is easy to imagine our state of paralysis if we had to be in a constant process of thinking consciously about each of our actions before moving or talking. When we speak, walk, or recall memories, a myriad of unconscious processes supports these activities. At the conscious level, we provide only broad strokes: the ideas we want to develop orally, the place we want to go to, or the episode of our life that we want to recall. Autopilots then do the work and produce a smooth and effective result. The whole of our mental life is thus articulated according to a permanent dialogue between two major systems: the conscious executive system and the system of habits (see Annex 7).

A drawback of these learned automatisms is that they put the brakes on change and innovation. Indeed, lasting change is only possible by opening another path, by tracing and wiring new circuits with regular practice. And the task is not easy, because it is difficult to rid ourselves of the consolidated circuits that already exist.

Our Brains are Biased

If we want to avoid, mitigate, or correct systematic inclinations that affect and distort our decisions and judgments, we must accept the fact that our brains are biased in some ways and learn how to define and name these predispositions. "There are some who would never have fallen in love if they had not heard there was such a thing".[4] If we are able to name our biases, we can then observe them and learn how to harness them. In fact, it is difficult to be aware of how biases affect our thinking as they are largely unconscious and automatic. They influence us, even if we do not believe it, just like optical illusions. So, to learn and recognize them, we must take a step aside and make a conscious effort to anticipate them in specific situations.

[4] François de La Rochefoucauld. *Maximes*. 1665.

We See the World in a Biased Way

We must accept the fact that our brains do not always operate in the rational way that we are led to believe, as they give us a subjective and somewhat transformed representation of the world. We are of course capable of incredible feats. We can speak our mother tongue with a great degree of precision, carry out complex mathematical or statistical reasoning, and immediately access a wide range of information. However, we perceive reality according to specific perspectives and transformations which are more potent than we imagine. It's like wearing glasses that distort or color our perceptions in certain situations, without us really being aware of it.

It should be noted, however, that what we perceive as biases today, in the environment we live in, may have been quite useful for survival in times which, in evolutionary terms, were not that long ago.

Overcoming Biases

We can't erase biases, but we can mitigate them by being aware of their influence, by naming them and considering them from different perspectives.

Overcoming Biases by Relying on Ourselves

In everyday life, biases can influence important choices such as getting married, looking for a job, or finding a place to live. For example, the bias of social conformity prompts most people to choose a spouse from a familiar environment or from a similar socio-cultural background.

By looking at how our choices affect us and those around us, or, by analyzing the decisions that did not produce the expected results, we can identify some strong trends in the way we behave. But this is not easy to do, as biases act surreptitiously, and overconfidence makes us feel that we are less biased than others. It is easier to see biases in others than in oneself. It is easier to see the mote in our neighbor's eye than the beam in our own!

In fact, a friend who spends thirty minutes with us may have a better understanding of our functioning than us, but there is often a conspiracy of silence around us. Our loved ones do not want to upset us, or spoil the time they spend with us, and those less close have other things to do, or prefer to leave it to others to pass on the message.

Overcoming Biases by Relying on Others

We need others to understand who we are, and we are very attentive to the way they observe us. We expect realistic and valuable feedback from advisors, coaches, or even colleagues and superiors if the culture is appropriate. Rapid and detailed feedback is therefore essential in daily practice, with the proviso that it must be handled with care, in the appropriate context. Our brain needs a fast feedback loop to learn because in a matter of days or weeks, most of the relevance of specific information fades away. In teamwork, it is even easier to become aware of and correct the biases of some team members, thanks to the diversity of points of view. Let us not forget, however, that the bias of social conformity may prevent some members from expressing a divergent point of view.

A Preventative Approach According to the Situation

Some situations, such as hiring a new candidate or making a major investment, may engender systematic behavioral biases which could be anticipated and mitigated by developing preventative measures, protocols, or decision guidelines such as: "if this…then". If such a situation arises, then do this and take that precaution. A link is established between some characteristics of the situation and the preventative action. For example, if we want to be ready for our next workout and not forget anything, we put our gear near the front door. This systematic approach ultimately leads to effective habits.

Mitigating the Seven Capital Biases

We have noted in the previous chapters seven biases that we consider as primary and dominant and that leaders should keep in mind. These biases are at the root of many other biases, and of many of our automatic behaviors, hence their designation as *capital biases*. However, they are not 'sins'—because they are involuntary as we usually are not aware of them.

These inclinations, deeply anchored in our brain circuits, now look like manufacturing defects in our present context. However, they must have been useful for our ancestors to solve survival and adaptation problems. They can still help sometimes but more often they are not appropriate or exaggerated. So, let's consider in turn the seven capital biases and see how to deal with them with some examples.

Four cardinal biases

Insecurity bias
Immediacy bias

Egocentrism bias
Hierarchical bias

Two process-based biases

Least effort bias
Confirmation bias

One social bias

Conformity bias

Bias of Fear and Psychological Insecurity

Fear speaks to our most primitive concerns. It has emerged from evolution to solve the recurring problem of survival in a dangerous environment and is now inscribed in our minds as a bias, a predisposition that makes us anxious to detect and anticipate any danger. The human brain is wired to overestimate and over-react to risk and threat before looking for rewards or anything else. Distrust is more natural than trust. The insecurity bias is always present during learning, which occurs mainly by trial and error. The fear circuit activates the amygdala and limits the normal functioning of our prefrontal cortex.

Fear remains pervasive in organizations creating a climate of insecurity. To avoid *being at fault*, many will deny their mistakes, hide them, or refuse to take initiatives that may go wrong. People avoid *speaking up* when they should, and they refrain from asking embarrassing questions when they consider the choices and decisions of their superiors. In fact, displeasing one's boss can weigh much heavier than seeking to achieve good results.

Preparing Well for Communication

The fear and risk aversion bias makes bad outweigh good and losses loom larger than gains. This distorted perception must be accounted for when communicating information. For example, if you are to announce poorer than expected results, you should word the message very carefully and bolster it with solid information and facts; otherwise it may quickly be interpreted more negatively than expected, morphing in people's minds into a threat of job losses and soon in massive layoffs. Journalists often play on this sensitivity by dramatizing events and emphasizing dangers that capture the imagination and fire up emotions, even when they are quite unlikely.

Developing a Culture of Psychological Safety

One way to counter the insecurity bias is to create a climate of psychological safety that allows everyone to feel secure and learn from their errors. People should feel confident to quickly spot, correct, or report failures. The issue is not with the failure, but with the fact that not reporting it prevents the correction of the process. So, employees are encouraged to speak up to signal problems.

Establishing this kind of culture requires a lot of effort and discipline over the long term. Well-intentioned leaders may approve of the notion of psychological safety, but they find it hard in the pressure of the moment to not raise their voices, frown, or display some authority. The culture of psychological safety is not decreed from above but carried out through a regular practice that actively rewards and protects candor and openness and celebrates those who raise interesting issues. It is this same concern that drives educators to encourage children, from an early age, to express their opinion, learn from their mistakes, or question the knowledge that some see as indisputable.

Bias of Immediate Reward and Future Discounting

The immediacy bias makes us want everything right away and to seek more than what we just experienced, always more, with the risk of addiction to immediate reward. This leads to shortsightedness and excess. Dopamine neurons are the basis of a very persistent incentive system, which continually seeks to experience more pleasure and will put in the time and effort required to achieve it. A serious design flaw, indeed!

When it becomes necessary to defer immediate rewards in favor of future ones, the urges of the *desire circuit* are inhibited by the *control circuit* which is connected to the prefrontal cortex. However, the activation of the control circuit is more costly in brain energy and more demanding than the activation of the desire circuit.

As gratification is delayed over time, future rewards are discounted. And this discounting effect is magnified in our society of impatience and quick fixes. When it is easy to have everything immediately, the ability to wait fades and the conscious activation of the prefrontal cortex to curb immediate impulses becomes ever more demanding.

Taking Some Distance to Curb Impulsivity

Before an important decision, routinely set a cooling-off period. This rule helps toward deciding with a cool head instead of getting carried away by the enthusiasm of some, or the fears of others.

Do not respond immediately to an email that is cognitively or emotionally loaded, sleep on it.

When filling a job vacancy, consider all possible candidates to avoid choosing among the closest and most immediate ones.

Deferring Gratifications with the Help of the Control Circuit

Cultural context and education are essential to internalizing, in the prefrontal cortex, and more specifically in the orbitofrontal cortex,[5] the representation of future gains in order to outweigh the immediate attraction of the short-term ones. For example, students can devote many years to preparing their career by depriving themselves of many immediate pleasures if they are able to imagine a promising future. The strategies put in place by some philosophical or spiritual movements, such as stoicism, essentially consist in trying to inhibit the short-term activity of the desire circuit through commandments or promises of future happiness or relief, arbitrated by the prefrontal cortex. But rather than trying to regulate behaviors, it is sometimes more efficient to rely on coercion, fear, and penance, or simply on automatisms and routines.

Using Coercion

Coercion may allow some control. Ulysses managed to resist the song of the sirens, ordering his crew to block their ears with wax and bind him to the ship's mast so that he could resist temptation. But let's not forget that as soon as the constraints disappear, people fall back into the old ways.

Internalizing Habits and Routines

It is also possible to use processes embedded in the subconscious in the form of habits and routines. It takes time to structure these processes and hardwire them through

---

[5] The orbitofrontal cortex is situated in the lower third of the prefrontal cortex, above the orbits. Please refer to Annexes 4 and 5.

practice and repetition. For example, soldiers in the special operations force in the Navy overcome the weakness of the will through intensive training (see Annex 7).

Egocentric Bias and Overconfidence

Being at the center of our attention, we strengthen our self-esteem by overestimating who we are, what we think, what we know, what we own, or what we do. We are also attracted to anything that can boost self-esteem. This may be professional success, compliments and applause, the quest for fame and adulation, or status from the purchase of luxury goods. We give more credit to our own beliefs and creations than to similar ideas generated by others. Competitors are thus regularly underestimated. This excess of self-confidence limits our ability to truly listen and accept feedback and evaluations from others.

Inflating our self-esteem and cosseting our narcissism provide pleasure and a release of dopamine in the reward circuit. As dopamine is at work, we tend to seek more stimulation against the background of habituation, and we may become dependent and addicted. But when the opposite happens: our dopamine level decreases, and we feel a lack of satisfaction which could be painful. Testosterone contributes to the egocentric bias by increasing self-confidence and driving people to be dominant, arrogant, and overly optimistic.

Deciding Who Speaks and When

In meetings, it is important to restrain some participants from dominating the debate and to regulate the overly assertive arguments of those who try to prevent dialogue. This can be done by setting up a process that allows everyone to fairly express themselves.

Remaining Modest

As the bias of self-centeredness is quite frequent among leaders, they must remember to remain modest in social interactions and take outside advice, even in areas where they think they excel.

Hierarchical Bias of Domination and Control

Like many other species, we have an implicit sense of hierarchy, which is based on thousands of years of genetic evolution and social conditioning developing from the need to cooperate in bands or tribes. Hierarchy signals position and status and engenders a classification system that formalizes unequal access to limited resources, be they monetary, physical, or status related. We continually emit waves of messages to recalculate our status in relation to others, and all the signals we receive, whether positive or negative, accumulate to define our position in the group. Complex neural maps in our brain represent our own status, and the hierarchical order of the people around us, in multiple dimensions.

The lure of social situations of domination and power continues to attract and fascinate. Successful leaders make great efforts to protect, strengthen, or expand their position and seek out potential allies to gain more power. Conversely, hierarchy can satisfy the need of some people for a disciplined world, and they accept to submit to this order.

Avoiding Triggering This Bias

Leaders can avoid triggering this bias in a meeting by remaining among other participants to signal that all voices will be heard. Likewise, neutral clothing like what is worn by the other members of the group is preferable. Asking other members of the group to present ideas and projects puts the ball at the center of the court. The organization and the management of the meeting could also be delegated.

Rotating and Redistributing Power. Fluidity of Roles

Occupying a leadership position in a group should mean that the leader has the necessary qualities to properly manage the group when required. This means that the leadership position could vary according to circumstances and needs, and that this position should not be given too much importance to facilitate the rotation of roles. Shifting positions in the group would also prevent some from getting intoxicated by the dopamine of power.

Giving up Status Symbols

As a leader, giving up status symbols is the most powerful message you can send to signal that you care about what your team members have to say.

Bias of Inertia and Least Effort

The brain is a *cognitive miser* that seeks to perform as few mental operations as possible for a satisfactory result. We are content to take the well-trodden paths to avoid recourse to slow and deliberative process of reasoning. This predisposition to *expedience* leads us, when we come to hesitate between what is easy and what is true, to choose, all too often, easiness. The effect of the least effort is that the desire to learn gives way to *lazy thinking*. The default choice or the status quo is privileged. In organizations, many people are content to obey, conform to the rules, and stick to the usual and reassuring way of doing things. "If it ain't broke, don't fix it".

Challenging Routines

The inertia bias prevents organizations from questioning their choices, for example, by keeping irrelevant businesses for too long, or by replicating the same budgets from one year to the next. Installing an anti-status quo routine enables fighting fire with fire: challenging the status quo becomes the default choice. Challenging routines becomes routine.

Halo Effect

The halo effect means that a single, easily noticeable, aspect of a person, is often used to make an overall judgment on that person. For example, in a recruitment decision you can easily be impressed by the size, look, expression, or diplomas of a person. To avoid being led by this tendency to least effort, it is important to complement the tendency to confirm a first impression by further cross-checking different sources of information.

Introducing Opportunities for Questioning

Automatic operations and habits are crucial for performance. However, care must be taken to avoid the cognitive comfort of routines and to introduce opportunities and pretexts for challenging the status quo.

Confirmation and Framing Bias

What we learn is inscribed in our brains in the form of neural networks, and any new information is connected to the neural circuits already present. Learning is thus done by integrating any new information with the networks in place, confirming what already exists and reinforcing our beliefs. Our brain is comforted with a burst of dopamine, a burst of pleasure when our beliefs are confirmed, and the reward from confirmation can become addictive. We keep going on well-trodden paths, making it difficult to unlearn and change. The consequence is that we tend to ignore or reject contradictory information. As a result, our beliefs and values are shaped by the way our experiences are averaged, summarized, and associated to previously stored information that move deeper into our subconscious, forming our identity.

Confirmation bias also applies to the way we perceive, interpret, and *frame* a situation. Given our narrow focus of awareness, a situation can only be viewed from a certain point of view, depending on the *focus of the attention.* Questioning is changing the framing, the point of view, by taking a helicopter view, or moving one step to the side. This is costly in terms of rational thinking and effort.

We are destabilized by *dissonant information* that confronts and challenges our thoughts and beliefs. This leads to a painful internal conflict called *cognitive dissonance.* It is difficult to hold together in mind two incompatible or contradictory views or beliefs. To mitigate this dissonance, we try to regain peace of mind by forgetting or denying one point of view and preserving the other with a reassuring interpretation based on all kinds of rationalizations or imaginary evidence. We also silence dissonance by closing the narrow door of awareness. Dissonance does not exist if we are not consciously aware of it.

Promoting Contradiction

This bias is corrected by introducing other perspectives to avoid over-reliance on established beliefs or thought. Hence the importance in scientific reasoning is of setting up experiments likely to falsify a given theory.

By welcoming and fostering contradiction, complacency and stubbornness are less likely to take hold. Thus, as people feel free to speak up and express themselves openly, it becomes possible to prevent some ideas from dominating the debate or reinforcing established beliefs. In fact, a single dissenter may be enough to break the spell of the status quo or preconceived ideas. But those who risk challenging the doxa often do so at their own peril if a benevolent climate of psychological safety has not been established beforehand.

It is also possible to facilitate the exchange of differing opinions by assigning specific roles to some members of the team. One person may play the role of the *devil's advocate,* and another could be the *fact checker*. It is also possible to have two teams working in parallel that defend conflicting or divergent points of view.

Reframing

Considering different points of view is cognitively costly. It requires trying to share the other's point of view or considering arguments that go against your ideas. One solution then consists of practicing experiments which force you to imagine situations where you must defend ideas or points of view that may seem, at first glance, totally incongruous or disturbing, but which would liberate you from the way you frame the situation. Changing the framework in which things are presented may reveal valuable discrepancies.

Breaking the Frames

Creativity can be simply defined as the ability to make unexpected predictions through improbable juxtapositions or analogies. To let our brain venture into more distant associations and establish new or unlikely connections, it is necessary to break the usual mental models by moving *out of known frames of reference.*

Let's look at the well-known puzzle where one must connect the 9 dots in a square with 4 straight lines, without lifting the pencil from the paper. Try to do it on Fig. 2, without looking at the solution below on Fig. 3.

The solution consists in getting out of the square delimited by the 9 dots and eventually of the plane on which the dots are printed to link all dots with a single line. You must be careful not to focus on the square delimited by the 9 dots, as you will remain stuck within it.

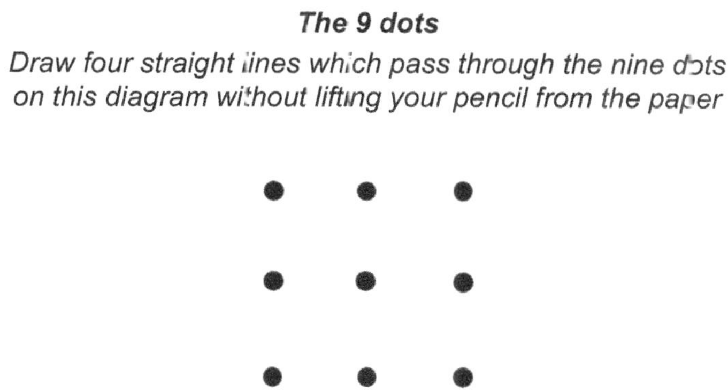

Fig. 2 Nine-dot puzzle

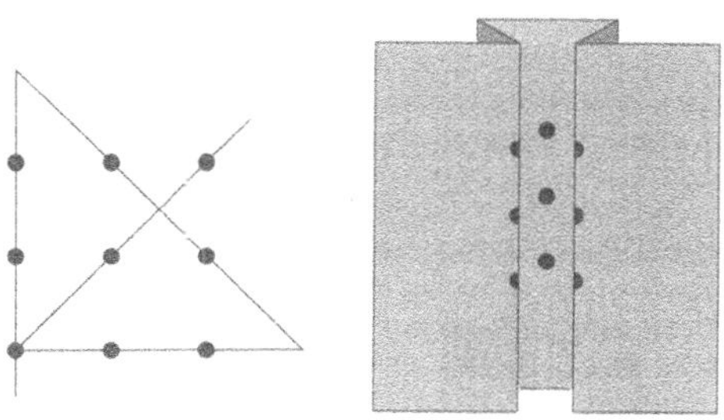

Fig. 3 Solution to the nine-dot puzzle

The prefrontal cortex is not the source of insight, and, even worse, it can act as an inhibitor. For insight to arise, you must stop reasoning and grasp subtle connections and weak associations.

Bias of Similarity and Social Conformity

Genetic and cultural evolution has favored socialization within the protective framework of the family or the tribal unit. The adaptive benefit of trusting community members was essential for comfort and protection. Familiarity and belonging to the same mental world cost less effort to relate to others and provides a feeling of security. As a result, we seek to associate with those who look like us, whether by profession, cultural habits, status, or language. When we are among peers, we share the same views, we forge strong bonds, and our identity is well established. We are thus conditioned to seek social acceptance, and we are willing to make a lot of efforts to be included in the group. A sense of belonging to a group is as basic a need as drinking or eating. Conversely, exclusion or ostracism can engender the same pain as thirst or hunger. Like other species we automatically distinguish between 'we', our loved ones, those from our side and 'them', the strangers, those from the other side.

The bias of social conformity leads to *groupthink* as individuals tend to belong and conform to the opinion of the group, to the majority, rather than risking open conflict. In the name of compliance, they can be led to accept absurdly incorrect propositions.

Choice Driven by Similarity

Have you ever chosen a candidate for a vacancy who comes from the same university as you? This bias, so frequently observed in daily life, could be offset by automatically widening the search. As Mark Twain explains "Whenever you find yourself on the side of the majority, it is time to pause and reflect".

Avoiding Easy Consensus

To prevent groupthink from returning insidiously when meetings are not well prepared, specific rules should ensure sufficient diversity in the group and encourage the expression of different points of view. It may also be relevant to refuse to take a consensus decision if enough valid alternatives have not been presented.

Groupthink can also be prevented with a *'premortem' approach.* The idea is to imagine that some time has passed after the decision has been taken and that the result is catastrophic. Each participant must then draft a short explanation for the disaster. This anticipated review helps to legitimize doubts and encourages the search for threats that have not yet been considered, or insufficiently discussed.[6]

Curbing the 'Us Against Them' Mentality

It is difficult to avoid the antagonism of the 'us against them' mentality, the mistrust of the other or the foreign group, short of 'destroying' the amygdala. It is therefore necessary to rely on the prefrontal cortex to try to overcome these reciprocal antipathies. This may be done by striving to consider these 'others' as individuals

[6] Klein (1998).

and not as a homogeneous entity steeped in stereotypes. Not everything would pit 'us' against 'them'. So, some options and some common objectives can enable a possible agreement.

Diversity and Inclusion

The more homogeneous teams are, the more at ease and confident they feel. But this feeling is misleading, as diversity increases overall team intelligence by combining multiple perspectives to solve complex problems, make good decisions, or innovate. The greater the diversity, the greater the potential value of exchanges in a system. Diversity therefore requires effort in the face of *cognitive laziness* and the *comfort of familiarity*.

Innovation requires the cooperation of different types of talents and skills. It is not achieved by bringing together similar people, however brilliant they may be. However, a diverse range of participants is not enough as these individuals need to cooperate and feel included in an active community with a common goal. And therein lies the paradox of diversity and inclusion.

A First Step but This is not Enough

The conclusion is that training should be organized to learn about the functioning of the brain and its limitations (see Annex 8). But making a personal effort to become aware of our biases and counter them is not enough. This effort should be integrated into a *systematic process* that will transform the organization and make it *a great place to work*.

Reference

Klein, G. (1998). *Sources of power. How people make decisions*. MIT Press.

Chapter 11
A Disciplined Process in Action

A Disciplined Approach to Generate Motivation and Swim Against the Tide

Leadership is not mere power-wielding and emerges from a *process* where leaders mobilize resources to motivate followers to collaborate toward a common goal in a supportive cultural context. Leaders must be ready to take the hard and long road that engages all actors.

But with elections around the corner or the publication of quarterly results, will leaders assume the long-term vision with determination? Will they give enough freedom to allow local autonomy? Will they promote trust and relational engagement? Will they encourage learning? Will they reward team collaboration and commitment to change?

These challenges require a disciplined approach based on five key factors to avoid falling back on the fixed model, which is the default model, and moving as close as possible to the inspiring creative model of leadership (Fig. 1).

In fact, leaders will make successive efforts on one factor or the other according to the situation while disciplining themselves to focus on the process, on the *how-to* go against their natural inclinations and biases and on the *how-to* energize and motivate people, before the *what-to do*, the performance. This means that the resulting model of leadership will evolve between the fixed and the creative model depending on the exercised determination and perseverance. If it is not possible to follow a disciplined process, the initiatives launched by the enthusiasm of a few will erode gradually, and the organization will backslide toward the default model. So, let's review what it takes to exercise this disciplined approach.

© The Author(s), under exclusive license to Springer Nature Singapore Pte Ltd. 2023

J. Teboul and P. Damier, *Neuroleadership*,
https://doi.org/10.1007/978-981-99-5122-2_11

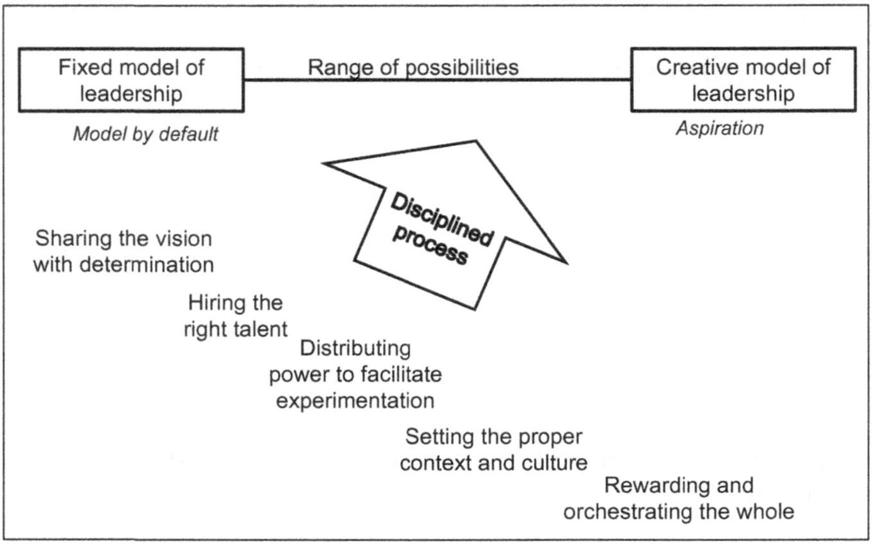

Fig. 1 A disciplined process

Sharing the Vision with Determination

A Clear and Meaningful Vision and Priorities

Any transformation program is at risk of getting bogged down without clear direction and long-term commitment. This does not mean that short-term results are not important. They must exist within the long-term framework. The strategic vision, analyzed rationally or inferred from acquired experience, is the backbone of a collective project that is spelled out into local concrete priorities in each unit or department. These few and well-chosen priorities allow everyone, at each level, to find meaning, understand the goal, and get mobilized. At Virginia Mason, Gary Kaplan aligned the whole organization to a common vision that led to the engagement of all stakeholders. This was not a foregone conclusion as he had to maintain the course with determination, despite multiple conflicts of interest.

Giving Meaning and Understanding the Why

Setting out a strategic vision is not enough as the hardest part is to ensure that the chosen direction will be understood and will give enough meaning and coherence to mobilize as many followers as possible. Clear and concrete messages should explain the vision and then be repeated tirelessly until they are internalized and approved to guide action, at all levels. The more these messages are simplified, teachable, and expressed in different forms and through multiple channels, the more easily they will be memorized and recalled. *Transparency of communication* will ensure that people will get access to the full information and can be trusted to receive it.

Determination from the Top

The efforts to put in place a creative model of leadership must come from a deliberate commitment of all leaders, starting from the top. At Virginia Mason, once the strategic direction was chosen after negotiation, officials at all levels had no choice but to commit to it. It is essential to start from the top because initiatives from lower levels are often stopped higher up by executives who are not fully convinced or by the difficulty of accessing key resources. Moreover, global change would not work on a piecemeal basis or in a disconnected way. So, Gary Kaplan worked hard to develop a close relationship with board members and key executives to get their commitment over the long run.

Hiring the Right Talent

Disciplining Behavior Through the Recruitment Process

The traditional recruiting system is very competitive as it pushes everyone to rise above the others to succeed. It focuses mostly on cognitive intelligence and technical or business expertise at the expense of social skills, thereby impeding the development of a culture of relational engagement.

It is therefore important during the recruitment process to verify that the selected candidates have a real motivation to learn and the social skills that will allow them to develop a fruitful collaboration. In fact, we know how difficult it is to change the orientation or the behavior of individuals as this means changing the way their brain is wired. So, it is important to get the right people with the right mindset from the start, when hiring. According to Laszlo Bock[1] VP of People Operations at Google at the time, "hiring is the single most important people activity in any organization… In addition to being willing to take longer, to wait for someone better than you, you also need managers to give up power when it comes to hiring".

In fact, at Google, they want to attract *'smart creatives'*, people who are conscientious but enjoy fun, team players with a creative energy and a learning mindset. They also believe that hiring should be peer-based, not hierarchical, with decisions

[1] Bock (2015).

made by committees. And once hired, employees should get opportunities to be constantly learning new things. And Laszlo Bock adds, "It's about finding the very best people who will be successful in the context of your organization, and who will make everyone around them more successful".

It is also crucial to identify the candidates who risk undermining the cultural harmony, those who have reneged on their commitments at some point or broken their promises, or those who show narcissistic tendencies. It is harder to be attentive to social predispositions and biases but betting on the wrong candidate could damage group cohesion as it is easier to destroy trust than to develop it. Bad behavior by a single individual can be contagious and affect the rest of the group. At Virginia Mason, once the agreement on the new vision was reached, all stakeholders were required to honor it and those who did not agree had to leave the organization.

Distributing Power to Facilitate Experimentation

A Reversal of Perspective

As they work to develop a creative model of leadership, leaders must operate a reversal of perspective and forgo the power struggles that have served them so well to reach the position they hold. They must now agree to redistribute some of this hard-won power, while disciplining their hierarchical and ego-centered predispositions.

As they empower their employees to take charge of their own operations, they help them audit their work processes and develop their skills. They will thus play the role of coach, mentor, or trainer as required. These roles require relational and teamwork competences that are as essential as technical knowledge and skills.

Individual selfish behaviors, political machinations, and struggles to reach the top will not disappear but will be seen as less beneficial and, above all, unrewarded in comparison to participation in the overall movement. It is no longer acceptable for internal competition and selfish behavior to destroy the culture of relational engagement and learning.

Ownership of the Work Process and Experimentation

As employees get better ownership of their work processes, they get more autonomy and control. They see and experience for themselves the advantage of collaborating with others. They now react less to vertical injunction and more to the reality of their situation, with the resources at their disposal and within the existing constraints. An informal culture of trust becomes established through regular practice and experimentation when individual-centered methods of evaluation and cumbersome reporting systems are de-emphasized, and accountability allows for reduced control.

Improvements and innovation will flourish around teams engaged in multiple projects. By proceeding step by step and retaining useful adaptations, these teams will learn to forget inadequate practices, correct their failures, and acquire the necessary competences.

Setting the Proper Context and Culture

Taking Power Away from Managers

In Chap. 9 we saw the importance of flattening hierarchies and reducing organizational complexity by having the problems addressed where and when they arise. As operators get visibility and control over their own processes, they can avoid higher level reporting and evaluation. Their authority is inscribed in the system, independently of the goodwill of superiors.

At Toyota, plant managers and team leaders are directly involved in operations as they have their offices located on the shop floor, and not in a separate location. Since operators are closest to the problem, they have the most useful information for coming up with a solution. Thus, decisions are taken at the most granular level of detail, without abstract data aggregation. The effects of each change are checked before its adoption, and this local, step-by-step experimentation prevents the unpredictable and amplified consequences of drastic changes and restructuring actions that blur the picture and destroy the culture.

At Google, a lot of power and authority is taken away from managers. Main decisions are either made by a group of peers or a committee. According to Eric Schmidt, the executive chairman at the time, "Managers serve the team", and Laslo Bock[2] confirms: "the default leadership style at Google is one where a manager focuses not on punishments or rewards but on clearing roadblocks and inspiring her team".

Reinforcing Transversal Cooperation

The adjustment effort does not remain at the level of work processes but must also encompass transversal cooperation, in line with the logic of customer demand and service orientation. The various units that follow each other horizontally along the value chain have to learn to cooperate quickly and well. And we know that this is not easy as each unit tends to develop a homogeneous culture that may be different from the next one. Here, again, we encounter the bias of social conformity and the tendency of self-contained groups to maintain their independence and compete,

[2] Bock (2015).

or even conflict with each other. The lack of cross-functional communication and integration is the root cause of waste and delays, employee anxiety, or customer anger.

It is futile to try to change behaviors through speeches or reasoning. Transversal cooperation does not get established by decree, exhortation, or constraint. People will find an interest in managing their interactions better if they have the incentives, the resources, and the motivation to do so. We illustrate this classic transversal dysfunction, with the example of a chain of hotels that had to deal with a poor room occupancy rate, and we show how managers resolved the problem by changing the incentives and redistributing power.

Hotel room occupancy

Managers of the chain of hotels in question had to deal with a poor room occupancy rate because receptionists were not renting rooms to guests who arrived after a certain hour.[3]

Initially, attention focused on the receptionists who did not seem to be acting in the correct way. Consequently, the way receptionists were evaluated was strengthened and management even initiated a specific training program, but to no avail.

On analysis, it turned out that complaints most often related to maintenance issues such as broken TVs, faulty faucets, or defective heating, leading receptionists to keep some rooms empty as back-up. The problem now stemmed from the fact that the hotel cleaning staff, fixated on their productivity goals, did not notice, or failed to report, some maintenance failures. Therefore, receptionists were dependent on the proper functioning of upstream cleaning and maintenance services, and the solution they found was to keep some rooms available in the event of unhappy customers.

The problem now was a lack of collaboration between the two units, and no longer the poor occupancy rate. So, to get the two departments to cooperate, management decided to give receptionists some control over the evaluation of cleaning and maintenance services. This introduced a relationship of dependence between receptionists and upstream services and a mutual interest in cooperating. Exhortation or training was not necessary as the new context made actors behave differently. Management aligned the organization with the final demand, by giving receptionists the role of *integrator* in direct contact with the end customer. The interdependence introduced by the redistribution of power benefited the entire team. As integrators, receptionists did not add complexity, as they were directly involved in action.

---

[3] Morieux et al. (2014).

Instituting a Culture of Relational Engagement

Offices should be designed to maximize energy and interactions. Employees do care about the place where they work, as illustrated by the way Google organizes its facilities with volleyball courts, climbing walls, gyms with personal trainers and lap pools, free gourmet cafeterias, and numerous kitchens.

The culture of relational engagement is built over time through multiple interactions that are regularly repeated day after day and which are then inscribed into increasingly implicit rules of conduct. These norms naturally develop through the sharing of experience and establish a climate of trust when behaviors become predictable and familiar. They become automatic and informal social shortcuts, like habits, and make interaction easier They encourage rapid collaboration and experimentation.

However, these norms must be followed and respected over time. The various partners are therefore invited to play according to the accepted rules and respond eagerly when someone asks for help. This mode of leadership therefore requires groups to function as integrated teams, testing consensus, and resolving conflicts. Meetings are also expected to reach effective decisions with the active contribution from all. If not, some discipline should be introduced, and the rule breakers should be sanctioned.

Ultimately, this effort to experiment and discipline behavior entails an *adjustment cost*. To collaborate closely with unpredictable partners, actors must make a significant effort to adjust their behavior instead of remaining distant and polite to avoid straining the relationship. So, what will convince actors to accept paying this adjustment cost? Here again, we find the need to establish a supportive culture supplemented by incentives and intrinsic motivators, such as getting proper recognition.

Instituting a Culture of Learning

For a culture of learning to take hold, we have already mentioned in Chap. 9 the importance of creating a climate of psychological safety which allows people to tolerate failure in order to learn faster. The correction of errors should focus on the process and not on people.

To see initiative flourish, people should be able to assert themselves and have the liberty to speak up when they think that an idea is valid—or when it is not. It must be safe to tell the truth and confront hard facts. Getting employees to voice ideas is an essential source of learning and progress.

Learning happens all the time and everywhere to develop and increase the level of competence with practice and experimentation. Leaders are there to help, guide, and educate.

Rewarding and Orchestrating the Whole

Rewarding Contributors but also Regulating Offenders

We have reviewed a number of motivators that leaders use to reward people and make the workplace a *Great Place to Work*. They vary according to circumstances, with *extrinsic motivators* such as monetary compensation, amenities, perks, or status, and *intrinsic ones* such as fun, recognition, freedom of action, relatedness, or fairness.

For example, people who respond positively to a request for help from a colleague should be singled out and acknowledged. And inversely not doing so or failing to honor a commitment would lead to a bad reputation and penalize future career development.

We have given a lot of attention to internal motivations, but extrinsic rewards and the associated release of dopamine are still very powerful. According to Eric Schmidt[4] the Google executive chairman at the time, "Once you get your smart creatives on board, you need to pay them; exceptional people deserve exceptional pay. Here again we can look to the sports world for guidance: Outstanding athletes get paid outstanding amounts". Nevertheless, there remains the danger of encouraging short-term focus and individual performance at the expense of teamwork and innovation.

Assessment of people's performance is difficult and takes time, persistence, and imagination. Managers in the operational field should be able to carry out this evaluation daily and give direct, regular feedback without waiting for the ritual annual performance review.

Disciplining behavior requires both rewarding those who accept collective teamwork and penalizing those who do not play the game, with graduated sanctions for individuals who cheat and seek power and influence at the expense of others. Often, peer pressure and good cooperation enable deviants to be identified and realigned without the need for sanctions. But it is essential to avoid deteriorating the existing culture under the influence of a few individuals.

Disciplining Behavior Through Education

Education prepares and unifies behaviors by promoting discussion, sharing the rules of the game, and establishing common practice. Training is much more effective when provided in small periods of a few hours that are time-spaced and supported by field practice. You may find it useful to refer to the ACES learning mode (Attention, Connection, Emotion, Spacing) presented in Annex 8.

---

[4] Schmidt (2014).

Managers have an essential role to play. They must be acquainted with the work of those they supervise and teach them how to master their work processes and solve their problems methodically. In so doing, they contribute to the development of human capital by cascading learning, teaching, and coaching.

Coherence. Orchestrating Initiatives

Realistic leadership must take a long-term view to consider the overall performance of the organization. The evolution of the whole is the result of multiple experimentations, repetitions, or contextual tinkering. Like natural selection, trial-and-error and multiple experiments lead to the emergence of systems that thrive in a complex ecosystem. This process is less easy to describe, less dramatic, but works better than centralized planning or big restructurings that shake up the whole organization.

However, complex systems cannot be managed by optimizing different parts separately, so it is important to have an *organizing committee* at the highest level responsible for the monitoring and orchestration of the whole process, as we have seen in the Toyota and Virginia Mason case studies. This structure supervises planned or unplanned variations and coordinates actions and methods, as well as reward and training systems, but it is not in charge of operations.

Leadership According to Jim Collins

Leadership in great companies according to Jim Collins

In his book 'Good to Great' published in 2001, after a decade of research, Jim Collins[5] explains how difficult it is to make good companies great. He points out that the successful transformation of an organization does not require *charismatic or big personalities* who make headlines and usually manage quick turnarounds by slashing the workforce and costs, reducing the R&D budget, and focusing on short-term results. As long as they remain at the helm, they can maintain some success, which however declines after they leave. Contrary to expectations, great leaders of big organizations are most often *unpretentious and modest.* They talk about the contributions of other executives and minimize their own role. In times of failure, they take the responsibility. But they show an intense and *unwavering professional resolve* while refusing mediocrity in

---

[5] James Collins. *Good to Great: Why Some Companies Make the Leap... and Others Don't*. 2001. The book was a bestseller, selling four million copies and going far beyond the traditional audience of business books.

any form. They first and foremost think about the success and the future of their organization.

With his '*hedgehog concept*[6]' Jim Collins highlights the importance of the *key strategic element* that leaders should consistently use to support every aspect of their decision making. This concept should be open enough to permit adaptation to different situations and the consideration of opportunities at all levels, all the way down to field operations.

Moreover, the framework presented by Jim Collins focuses on *discipline*, disciplined people, disciplined thought, and disciplined action. We are not far from the disciplined process that we recommend.

Disciplined People

Hiring good talent. 'First who, then what'. "Getting the right people on the bus and the wrong people off the bus". "Nothing is more important to a business than the talent it employs".

Disciplined Action

It's about practice and experimentation. Leaders are led by a constant drive to make steady, consistent progress toward their goals, with restless curiosity. Transformation does not happen in one fell swoop. There is no single big action or killer innovation. Instead, it is more like relentlessly pushing a giant, heavy flywheel, to gradually build momentum, step by step. With disciplined action we are close to the idea of continuous improvement and kaizen. Similarly, the creative model of leadership that we propose is based on a continuous and disciplined improvement approach.

Disciplined Thought

Disciplined thought deals with the learning culture. Leaders need to *confront the brutal facts*, as they are often in denial about what is going on around them. They must create a climate where truth is heard from their own people who are allowed to speak up. They should lead with questions and engage in dialog and debate, without using coercion. They conduct autopsies and analyze errors without blaming anyone for it.

[6] Berlin (1953). A fox knows many things, but a hedgehog knows one big thing. Hedgehogs interpret everything in the light of some single all-embracing system, a single pattern that they believe in and adopt. Foxes have a more pluralistic approach and can take many directions.

Netflix's Innovative Approach

We have looked at examples of organizations in the industrial or information technology sectors and in hospital professional services. Let's now consider an example drawn from the creative economy, which is growing in importance. While the Netflix case is extreme in some respects, it can highlight the main aspects of the disciplined approach we have just reviewed.

Netflix is an entertainment company that has moved step by step from a postal DVD rental system to an internet streaming service and now to a fully fledged integrated company producing its own content, its own movies and TV shows, with over 167 million subscribers in 190 countries.[7]

The Strategic Approach

Reed Hastings, the CEO, quickly realized the strategic importance of moving the company from being a distribution channel to a content creator. He also realized that what matters above all, in the entertainment sector, is innovation, novelty, and speed of reaction—much like in the fashion industry.

He started from the idea that the best employee in a creative role is ten times, if not a hundred times, better than the average employee. The best creative advertising expert can create an ad that will attract millions more customers than the average one. So, the model that best explains Reed's approach is the model of a professional sports team. Players of a baseball or soccer team are highly talented champions, and their success is based on a *'high density of talent'*. They are very close to one another, and they know each other so well that they can easily anticipate their next moves.

A Deliberate Strategic Commitment

Reed Hastings set out to systematically transform his organization, step by step, with a compelling and disciplined resolve. The gradual and orderly evolution in the same direction over many years has allowed the company to grow and adapt in a constantly evolving world. This example is particularly illustrative because it highlights the main points underlined in this chapter: the CEO's unwavering commitment to a perfectly transparent strategy, the freedom of action based on responsibility and discipline, and the systematic experimentation toward developing a collaborative and learning culture in every corner of the organization.

---

[7] Hastings and Meyer (2020).

Sharing the Vision and Full Transparency of Communication

The more employees at all levels understand the strategy, the financial situation, and the context of what is happening, the better they can make responsible decisions without the need to involve their superiors. And to build trust, nothing should be kept secret, whether it's a matter of reorganization, layoff, or financial and strategic information. This transparency is only possible because the organization is made up of motivated and self-disciplined people who need all the information available to fully do their job and who can be trusted not to disclose it.

Who? Hiring the Right Talent

A fundamental element of Netflix's approach is the decision made from the beginning to recruit only the best creative talents possible and to put them in small, highly integrated teams. In these teams, they work with amazing colleagues, from various disciplines and different backgrounds, who are exceptionally competent and innovative, but also able to collaborate effectively.

Increasing the Density of Talent

By strengthening the 'talent density', the company can attract the best performers who learn more from each other and thus work better and faster. It should be noted that these individuals have high levels of self-confidence, and for them, job security is not a priority. They are ready to forget what they have learned, or surpass it, to reinvent themselves.

Highest Pay but no Performance Bonuses

Hiring only the best creative talents leads to paying them at the top of the market. What matters is to give a substantial salary without introducing a *pay-per-performance bonus*. In Netflix's view, salaries are based on the market and performance bonuses do not help innovation because they encourage short-term focus.

Having the Best Person in Each Position

However attentive one is, recruitment errors are possible—or the company may need to change orientation in some respect. To maintain the highest density of talent possible, the company must ensure it has the best person in each position and thus take the difficult decision to part with good employees when they can be replaced with better ones. There is always the question of retaining a person who is considering leaving. Are you ready to fight to keep a top player in the team? If you are not sure of the answer, you must give this person clear feedback on how to move forward. As a result, some employees are forced to leave the company, but prior to this, they would have heard clearly and regularly what they need to do to improve. In the event of departure, they receive generous compensation that allows them to prepare for their next career step, and every employee who agrees to join Netflix subscribes to this approach. They agree to receive candid feedback on how to improve their performance and to be replaced if need be.

Distributing Power and Facilitating Practice and Experimentation

Don't Seek to Please Your Boss

Since most of the activity is based on the ability to innovate or do one's job creatively, it is essential that local managers retain their freedom of action and are guided by the unconventional instruction 'Don't seek to please your boss, seek to do what is best for the company'. These managers are provided with all possible information to enable them to implement the brilliant ideas they believe in, without close supervision.

Autonomy and Freedom of Action. Some Rules

As they gain more autonomy, employees feel more motivated to engage in their work and make the company more agile and flexible. Freedom of action operates against the hierarchical bias but is likely to be derailed without the responsibility and the informal trust that go with it.

So, at Netflix there is no need to get decisions approved and *'informed captains'* can sign their own contracts. They must look for what is good for the company, without necessarily trying to please their superiors. They will not put aside a great idea because their boss may not approve of it or does not agree with their views. Nevertheless, it should be noted that this freedom of action is partly due to the fact that the company is in the creative entertainment market and not in an area where risk can be fatal, such as medicine or nuclear energy.

However, some rules must be followed and respected. An informed captain who passionately believes in her project will only stake her bet after having 'socialized' her idea and farmed for dissent in many ways. She will then have to test out the idea. If successful, she can legitimately celebrate, and if she fails, she will have to make it known publicly and learn from it. Learning is facilitated by a culture of trust that requires everyone to express their disagreement if they are not in tune with an idea. They are considered disloyal to the company if they refrain from speaking up when they have a different opinion or have some feedback that could be helpful to a colleague. This is a way to fight against the bias of social conformity that prevents people from expressing dissenting opinions.

Practicing the Proper Context and Culture

The Strategy Spreads Out like a Growing Tree

The strategy that defines the contextual situation develops like a tree that unfurls and spreads out based on opportunities. This is made possible because Netflix is a weakly coupled organization unlike highly integrated traditional companies, such as Toyota, where all elements are so connected that a change in one part of the system can have implications on other parts. In contrast, when the system is weakly coupled, there are few interdependencies between the different parts, and you can modify one without noticeably disturbing the whole system.

So, at Netflix, if an opportunity does not turn out as expected, it does not have much effect on the rest of the organization. The company can therefore deploy its strategy like a growing tree, the reverse of the traditional pyramid where decisions move top down. Reed Hastings is at the bottom, at the root level. He pushes the company to grow globally and holds up the 'trunk' of senior managers. The latter defines levels of engagement and risk and supports external branches where decisions (the tree's fruits) are made by the informed captains. By giving them a great deal of freedom, this agile approach increases the flexibility of operations and accelerates decision making.

The main risk of weak coupling (low level of connection between parts) is the loss of alignment between the interests of the different stakeholders. Major decisions taken at the individual level must result from a close agreement between the boss and the collaborators who share the same vision.

Consolidating a Culture of Relationship Engagement and Learning

Netflix assumes that what matters to top performers is the joy of being surrounded by people who are both highly competent and able to collaborate, people who can help them progress. To be excellent is not just accomplishing one's task remarkably, it's also learning to collaborate by being generous, knowing when to pass the ball and putting the team before one's ego, in order to win together.

A culture of candor and relational engagement emerges when employees agree to participate in regular feedback loops that ensure that their actions are aligned with the good of the company. This reduces the need to resort to traditional controls.

The rule is to say what you really think, with a positive intention, without attacking or hurting anyone, and putting your feelings and comments on the table to deal with them. This positive feedback must be actionable and help recipients improve what they do. The latter must react appreciatively, without finding excuses. For these rules to work effectively, Netflix does not hesitate to get rid of those who have trouble identifying the boundaries of balanced behavior, however brilliant they may be. They risk damaging the culture, and this is especially true of people who think they are superior to others or who have narcissistic personalities.

Fighting Against the Bias of Fear and Psychological Insecurity

A candid culture requires people to accept remarks and possibly criticism about their work without raising doubts about themselves or their competences. This kind of feedback, in a less auspicious environment, could be threatening to most people, due to the bias of psychological insecurity.

Likewise, having the courage to openly give feedback to peers or superiors can generate legitimate concerns. Is my boss going to be mad at me and hold this against me? Isn't it going to hurt my career? Managers should show signs of reassurance in the form of appreciative gestures and public thanks.

A Learning Culture: Accept Your Failure and Make It Public

Once your competence is established, you can speak openly and extensively about your own mistakes. If all leaders do the same, trust and goodwill will spread throughout the organization. In fact, the greatest benefit of candor emerges from the moment employees begin to give real feedback to their superiors. The latter must show that it is safe to do so, by responding with gratitude to all criticism and above all, by giving signs of reassurance and belonging to the team. They can then share the evaluation of their actions with their teams by candidly detailing what they did wrong.

A Culture of Commitment and Candor Reduces Controls

As the 'talent density' increased, and feedback became more frequent and candid, it was possible to remove many authorization and approval systems across the organization. Netflix began by reducing controls on travel expenses, then continued with leave policies, proceeding step by step, procedure after procedure. Phasing out controls that stifle innovation speeds up operations and allows the company to function without too many regulations piling up.

Orchestrating Initiatives and Swimming Against the Tide

Rewarding Contributions but also Regulating Offenders

Expense claims are not audited, but the financial department audits 10% of expenses annually. If there is the slightest issue, the employee is immediately dismissed. There is no second chance: "Abuse the freedom and you are out!" This message is given loud and clear, and it serves as an example. Without it, freedom does not work. Managers must have the courage and discipline to terminate employees who display undesirable behavior or misbehave. Bad behavior is contagious and could bring down everyone's performance.

An Integrated Approach

This approach, having evolved gradually, now constitutes a whole. You cannot take one element without the rest. For the whole to function harmoniously, all the protagonists must agree to a tireless combat against any of the biases we have described. Accepting the psychological insecurity created by density of talent, accepting deliberate long-term engagement while refusing the temptation of immediate rewards, developing a culture of cooperation and relational commitment by rejecting egocentrism, reversing the hierarchical pyramid by freeing decision making, developing creativity and innovation against inertia and confirmation bias, agreeing to give feedback, and expressing dissenting opinions to escape social conformity.

It is an ambitious program that the company will have to keep up with discipline to avoid regression.

References

Bock, L. (2015). *Work rules! insights from inside google that will transform how you live and lead.* John Murray.

Berlin, I. (1953). *The hedgehog and the fox: An essay on Tolstoy's view of history.* Weidenfeld and Nicolson.

Hastings, R., Meyer, E. (2020) *No rules. Netflix and the culture of reinvention.* Random house.

Morieux, Y., Tollman, P., & Beslon, C. (2014). Smart simplicity. Manitoba Les Belles Lettres.

Schmidt, E. (2014). How google works. John Murray Press.

Chapter 12
Conclusion. Keeping the Human Side in Mind

We have been keen to stress the importance of associating leadership with followership to show the evolution of the notion of leadership from the vertical and fixed approach to a more disciplined and creative one. The traditional command-and-control system remains the easy path that leaders follow when they are driven by their natural predispositions and when motivating and mobilizing people remains problematic.

Our journey has led us to describe two models that served as reference points, at the two ends of the range of most observed practices. The fixed model is the one on which leaders easily fall back when they just exercise power over people who remain passive and obey. At the other end, the creative model mobilizes actors to cooperate and innovate when leaders follow a much harder and disciplined process that motivates everyone and creates the proper cultural context.

Two Benchmark Leadership Models

The Fixed Default Model

The fixed model is the default model on which leaders fall back when they don't invest enough efforts to inspire people with a convincing strategic vision, an encouraging leadership style, or a stimulating cultural context.

This model has a Taylorian flavor based on the rational and methodical approach inherited from the industrial revolution and the standardization of operations. The cold rationalism of the bureaucratic pyramid encases individuals within a rigid framework and well-defined functions. The model worked well because it was based on a mode of command-and-control, easy to operate with little consideration for people's motivations or predispositions. This stable model provided some protection

© The Author(s), under exclusive license to Springer Nature Singapore Pte Ltd. 2023

J. Teboul and P. Damier, *Neuroleadership*,
https://doi.org/10.1007/978-981-99-5122-2_12

and potentially permanent employment but showed a lack of flexibility and innovation. It proved painful in terms of individual freedom and produced stress and demotivation at work.

Nevertheless, it is not uncommon for employees, who are struggling with the new economy and find difficult adapting to rapid change, to have a certain nostalgia for the bureaucratic organization despite its rigid aspect. In the end, the system leaves some room for maneuvering at intermediate levels as power is diluted all the way down the pyramid.

In any case, leaders must work hard to avoid drifting back to the fixed model. Strategic thinking bows to the imperative of immediacy and short-term results. Vertical leadership reasserts itself when internal rivalry and power struggle undo cooperation. And when leaders strive to empower people, their efforts are hindered by the development of digital information which favors centralization. With the rapid change of context, the culture of relational engagement does not have enough time to roll out and a formal transactional culture holds sway, hindering trust. The insecurity bias hampers the learning culture, while the threat of globalization and the expansion of new technologies leave the employees who cannot easily adapt on the sidelines.

Nevertheless, some leaders succeed in moving away from the fixed model and improve the situation by playing on one, two, or the three dimensions that we have mentioned. They must engage, with discipline, on a harder pathway to motivate and mobilize followers. But when their efforts are not successful, they end up accepting the shortcomings of the default model with a sense of fatalism. When the sociologist François Dupuy describes these challenges, his books have evocative titles such as *'Lost in management'*, *'The failure of managerial thought'*, or *'The fatigue of the elites'*.

The Creative Model

The easiest way to get away from the default model is to have the whole organization aligned on the vision of a charismatic leader who represents the aspirations of the group. Engagement, motivation, and enthusiasm are present from the start as the leader is carried forward by followers. So, collaboration and learning naturally follow. But given the complex and unstable environment we live in, this alignment does not happen often or is difficult to maintain, and leaders must accept to embark on a long and disciplined process that will mobilize followers to collaborate and learn.

We have represented this trend with a specific model, the creative leadership model, a three-step process which makes explicit three key aspects of leadership: strategic thinking, leading style, and contextual culture. However, this ideal model often remains a mirage if leaders are not determined to face the complexity of a volatile environment and are not prepared to go against the tide of their own predispositions with application while mobilizing their followers.

A favorable ecosystem develops when modest and dedicated leaders strive to transform the context and the culture according to the disciplined practice that we have described. If they can demonstrate a brilliant and charismatic leadership style, so much the better, but it is essential that the learning organization moves and adapts, step by step, drawing on the efforts of all. The global result can be measured by an increase of *social capital* which represents the quality of the engagement and bonds that unite its members.

This favorable ecosystem establishes norms that make it easier to exchange ideas and connect brains, skills, resources, or concepts. But virtuous practices and culture erode. Trust and norms eventually fade away under the influence of our natural predispositions. Again, deliberate efforts are necessary to keep the ecosystem alive.

The problem with traditional leadership training seminars is that they seek to convince participants that the implementation of the proposed prescriptive model is possible if only they would agree to reform themselves. These efforts are misguided when they remain at the level of individual advice, wishful thinking, beautifully reinterpreted stories, or charismatic heroes. Once they return to their institutions, participants encounter a context that has not changed, and soon have no option but to resume their usual behaviors. They do what they can in the situation where they are placed, and the resulting disappointment leaves them even more frustrated.

The Importance of Knowing How the Brain Functions

Leadership training cannot ignore the functioning of the brain and its flaws. The more we believe that we are rational beings, firmly installed at the controls, the more we tend to remain naive, hostages to immediacy, prisoners of our fears, our beliefs, or the dominant social norms. Our education and upbringing have not taught us to welcome our emotional intelligence or recognize our natural inclinations and biases. It is about time we acknowledge them in our daily behaviors.

The way our brain learns suggests abandoning information-dense seminars cramming in too many concepts, because too much information kills information. It is better to move toward learning methods that favor just a few concepts that can be quickly presented in an hour or two, followed by implementation with concrete examples and projects shared by the group and management. The advantage of this sharing is to instill new behaviors through regular practice and to allow people to get to know each other more intimately by working together. This learning approach will create an informal culture that facilitates exchange and trust.

Putting Relational Culture Back at the Heart of Things

It is time to give collaborative and creative leadership the place it deserves beyond the traditional fixed model of leadership. This approach is not sufficiently practiced in the rational and technical world we work in, particularly in Europe, still dominated by Cartesian thinking. There is no happy leadership unless we are determined to reach it through collaboration, with application and discipline. To be able to evolve and survive with agility in a volatile and unstable world, complex organizations must be able to create the conditions that will motivate and mobilize all employees to think, imagine, decide, and learn together, toward a common goal. Cooperation between humans is born from the tension between the desire for freedom and the need to forge links. And this dual aspiration is best expressed when the environment allows for experimentation and learning, and the culture establishes relational engagement over time.

Annex 1—A Handbook

Some Information About the Brain

A brain weighing about 3lbs (1300 g), contains nearly 100 billion neurons (86 billion, to be more precise). With up to an average of 10,000 synaptic connections between neurons, we have close to 10^{15} or almost a quadrillion connection (like the memory of the most powerful computers).

With a moderate consumption of just about 30 W, the brain works 24 h a day, 7 days a week, 365 days a year.

Representing only 2% of our body mass, our brain expends 25% of the energy our body consumes at rest. So, it seems that we live in order to feed our brains. This means that the metabolic cost is 500 kcals out of the 2000 kcals we need each day.

Main Regions of the Human Brain at a Glance

- The dorsolateral prefrontal cortex (DL-PFC): **the reasoner**
- The orbitofrontal cortex (OFC): **the emotional regulator**. Its ventromedial part serves as **the evaluator.**
- The anterior cingulate cortex (ACC): **the controller** (also the conflict detector)
- The basal ganglia (central gray nuclei): **the curator**
- The amygdala: **the sentinel**
- The ventral striatum: **the rewarder**
- The hippocampus: **the memory explicitor**
- The hypothalamus: **the activator**
- The insula: **the body sensor**
- The ventral tegmental area (VTA): **the sprinkler**

© The Editor(s) (if applicable) and The Author(s), under exclusive license to Springer Nature Singapore Pte Ltd 2023

J. Teboul and P. Damier, *Neuroleadership*,
https://doi.org/10.1007/978-981-99-5122-2

The Cast of neuromediators[1] *at a Glance*

- Dopamine: **the stimulator**
- Serotonin: **the mood regulator**
- Noradrenaline: **the over-amplifier**
- Acetylcholine: **the learner**
- Cortisol: **the gladiator**
- Testosterone: **the dominator**
- Oxytocin: **the pacifier**

NB: Summarizing brain functions as a juxtaposition of independent centers, and neurotransmitters as simple functions, is reductive, but helps clarifying things for those not expert in neurosciences. The brain works through large networks of multi-connected neurons with regions that play a particularly important role in certain functions (a bit like airport hubs). Neuromediators have multiple actions that can also vary depending on the region in which they are present and the context.

The Seven Capital Biases

Four cardinal biases

Bias of fear and psychological insecurity. ***Insecurity bias***
Bias of immediate gratification and future discounting. ***Immediacy bias***
Egocentric bias and overconfidence. ***Egocentric bias***
Hierarchical bias of domination and control. ***Hierarchical bias***

Two process-based biases

Bias of inertia and least effort. ***Least effort bias***
Confirmation and framing bias. ***Confirmation bias***

One social bias

Bias of similarity and social conformity. ***Conformity bias***

Other Biases

More than a hundred biases have been described. For the sake of simplicity, we have isolated the seven biases that we think are particularly important in the field of leadership. Below we describe two other biases to which we referred in this book.

[1] Neuromediator is a generic term of representing neurotransmitters, neuromodulators, and hormones.

Attribution Bias

We are born with a propensity to make intentional attributions, to seek a purpose, an intention. We saw that we systematically tend to attribute the success or the failure of an organization to its leader.

When somebody does not do something well, you say 'this guy is clumsy' before considering the situation and the difficulty or the complexity of the task. When someone walks past you and does not greet you, you immediately think 'She's ignoring me' or 'She doesn't find me interesting', instead of considering many more trivial reasons that can explain this oversight.

When the information we expect is incomplete, we are inclined to think that the cause is intentional, even malicious, rather than accidental or the result of chance. We can expect business leaders and policy makers to attribute the success of their initiatives to their extraordinary abilities, and not to chance or circumstances.

Bias of Emotional Reasoning (also Called Affective Realism)

This is reasoning from what one feels, without supporting this reasoning with evidence or facts. We clothe feelings, whose origins we are unaware of, in the garb of reason and rely on this rational appearance.

We tend to believe our emotional perception first and foremost. What we feel when we experience an event can distort objective reality and influence what we believe. What we like determines what is valid. We follow our heart: "Everything I feel to be good, is good; everything I feel as bad, is bad". "I feel angry, it must be your fault".

Annex 2—The Big Five Personality Traits

The 'Big Five' Model

The 'Big Five' or the 'Five Factors Model' refers to five broad dimensions that are used to describe human personality. The model was defined by Lewis Goldberg, and developed by Paul Costa and Robert McCrae, using factorial analysis from the thousands of words and traits that describe personality in spoken language.

The five main dimensions are: Openness to experience, Conscientiousness, Extraversion, Agreeableness, and Neuroticism (OCEAN).

Introducing a Sixth Factor, Assertiveness

We think it is useful to set apart a sixth factor, ***assertiveness*** (self-assertion) by separating this trait from the extraversion factor. Extraversion would then correspond simply to the orientation toward the outside world (as described in the MBTI, Myers Briggs Type Indicator).

We have thus organized the six factors into two groups of three factors each, a group that reflects ***stability*** and a group that expresses ***plasticity.***

Meta-Trait: Stability

So, three factors are grouped into a ***meta-trait*** that corresponds to some positive or negative forms of stability: Neuroticism-Agreeableness-Conscientiousness. A number of studies indicate a strong link with ***serotonin, the mood regulator***. Low levels of serotonin are associated with aggression, poor impulse control, and depression.

© The Editor(s) (if applicable) and The Author(s), under exclusive license to Springer Nature Singapore Pte Ltd 2023

J. Teboul and P. Damier, *Neuroleadership*,
https://doi.org/10.1007/978-981-99-5122-2

Neuroticism: Withdrawal and Volatility Versus Emotional Stability

To understand neuroticism better, it is useful to contrast it with emotional stability.

Are you easily riled or calm? insecure or secure? Anxious or emotionally stable?

Emotional stability is the kind of equilibrium we experience when we feel alive and at ease with a good emotional balance. As we lose balance and flexibility, we move toward a state of chaos, with excessive arousal, or toward a state of rigidity or depression.

The amygdala plays an important role here because some people are hypersensitive to the slightest negative signal and have difficulty coping with it. They detect, recall, and ruminate over perceived threats or dangers that a more stable person would not notice.

It seems that low levels of serotonin and noradrenalin—two neuromediators that modulate the system that inhibits behavior—can lead to a lack of control and restraint.

Agreeableness: Compassion and Politeness

At the opposite of agreeableness, we find a critical, distant, distrustful approach. These two traits are at the two ends of a continuous spectrum.

Are you trusting, looking for peaceful relationships, or are you critical, objective, and suspicious? Do you tend to be helpful, or do you favor your own interests?

This dimension corresponds to altruism and pleasant co-operation, as opposed to cold and calculating behavior. It is clearly connected to the need to belong and be included in the group. ***Oxytocin, the pacifier,*** seems to play an important role.

Conscientiousness: Industriousness and Orderliness

Conscientiousness can be viewed as the opposite of irresolute and disorganized at the other end of the spectrum.

Are you organized or spontaneous? Self-disciplined and assiduous or indecisive and unmethodical? Do you feel a keen need to succeed in your projects or not?

This dimension refers to our aptitude in controlling our behavior to follow rules and pursue long-term objectives. We have seen that self-control uses the brain brakes of the lateral prefrontal cortex. But more generally, we must trust our prefrontal cortex and working memory to plan action and follow complex rules. Nevertheless, spontaneous, impulsive behavior can be advantageous when it becomes necessary to deviate from routines and change direction in a disruptive and rapidly changing environment.

Meta-Trait: Plasticity

We have regrouped the last two remaining factors, plus assertiveness,[2] into a ***meta-trait*** which represents a certain form of ***plasticity*** with a very strong link to ***dopamine, the stimulator***. This is in line with the exploratory tendency of dopamine that steers us toward rewards and novelty, and the cognitive predispositions of the prefrontal cortex.

Extraversion: Enthusiasm (Dissociated from Assertiveness)

Extraversion versus introversion are traits that are often cited in studies of personality.

Are you sociable, able to connect easily with others, or instead distant, preferring to observe rather than interact? Are you fun-loving or reserved?

This factor reflects the need to explore the world in search of exciting sensations, a search for rewards and novelty associated with dopamine. By contrast, introverts are less direct and intense in their interactions with the external world. They are more nuanced when they express their opinions, and they prefer to maintain some distance from others. Their energy comes from inside.

Openness to Experience: Openness and Intellect

Here, openness and intellect are contrasted with conformity and concrete thinking.

Are you imaginative or down to earth? Independent or conforming? Are you easily bored with material and repetitive tasks or at ease with concrete and practical things?

This dimension reflects openness to experience and the expression of sensory and aesthetic feelings. The person's intellect is easily engaged in exploring and appreciating new ideas and unusual creations. This often leads to refusal to conform to the group and its single way of thinking. In contrast, conformity refers to the desire to remain nicely installed in the comfort of routines without having to confront novelty.

This factor can be associated with the workings of attention and the complex information processing of the prefrontal cortex and the frontal pole in charge of abstract processing and the integration of multiple cognitive operations

[2] Extraversion in the 'Big Five' is composed of two traits, enthusiasm and assertiveness. We have removed assertiveness from extraversion which now mainly corresponds to enthusiasm.

Assertiveness and Self-confidence

Self-confidence and self-assertion are at odds with lack of self-confidence and the need for approval.

Do you need others' approval to give you permission to act? Are you decisive and determined in your decisions? Do you feel the need to prove your worth? Do you feel that being accepted by others is not a given? Do you feel that you normally find your place in a group?

We think it is useful to separate this trait from extraversion to avoid muddying the notion of extraversion as it is presented in many personality tests. One can be introverted and assertive, equally, one can be extraverted and lack self-confidence.

Annex 3—We Are Our Synapses

The basic unit of the neural system is the ***synapse***, the tiny terminal junction between neurons where an electric signal gives rise to an electrical transmission, a chemical transmission or both. Through the release of neuromediators[3] in the intercellular space, the synapse plays a crucial role in modulating our interior states, drives, and emotions.

Neurons and Synapses

The brain is a massive organization of complex and densely packed cortical networks containing almost 100 billion neurons, 600 million of them projecting from one hemisphere to the other in the corpus callosum. It is interesting to note that in our intestines we find a more modest network of about 100 million neurons to run the chemical refinery of our 'gut brain' with a high degree of autonomy.

A neuron can produce a voltage like that of a battery and generate invariant electric pulses called ***action potentials***. It is the ***variation in the frequency*** of these action potentials that carries information within neural networks.

[3] Neuromediator is a generic term of representing neurotransmitters, neuromodulators, and hormones.

© The Editor(s) (if applicable) and The Author(s), under exclusive license to Springer Nature Singapore Pte Ltd. 2023

J. Teboul and P. Damier, *Neuroleadership*,
https://doi.org/10.1007/978-981-99-5122-2

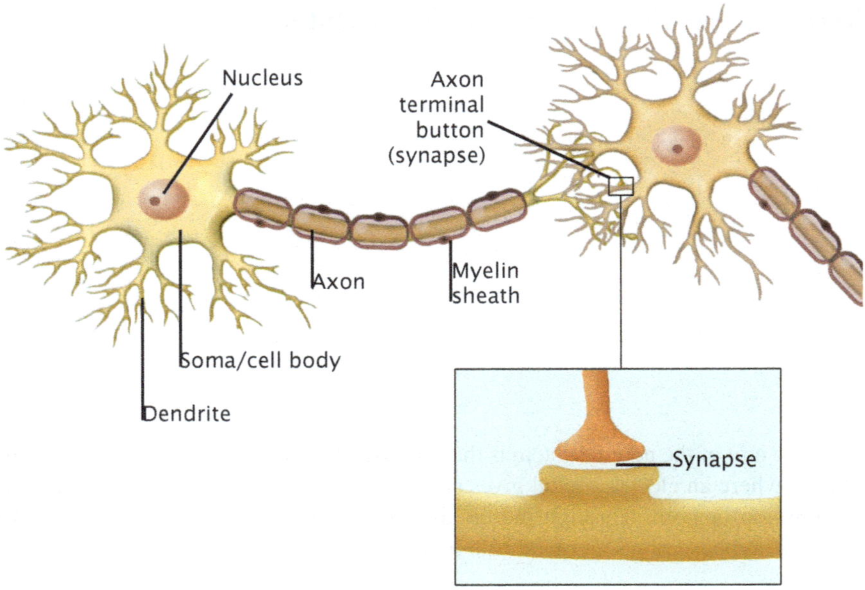

Fig. 1 Neuron

In the brain, neurons are embedded in a supporting structure made up of ***glial cells*** that provide them with energy and oxygen. Each neuron is composed of three elements: the cell body[4] (with nucleus, cytoplasm, and mitochondria), the axon, and the dendrites that extend out from the cell body and terminate in synapses. Each neuron has a large number of dendrites and up to an average of 10,000 terminal synapses.

Each neuron functions as a unit of integration that receives electric inputs from thousands of other neurons via a cluster of dendrites and sends its output by way of the axon, which can measure up to one meter for some cells. The electric signal (action potential) travels along the axon until it reaches a point of contact with another neuron through tiny terminal junctions, the ***synapses***. Then the communication between neurons takes a chemical form, with chemical exchanges at the junction (see Fig. 1).

Communication and Memory

Each synapse has a double function. It is both the point of communication between neural cells and the site of memory storage and learning.

[4] The nucleus contains the DNA, support of our genome, the cytoplasm contains all the machinery to produce protein and other chemical components, and mitochondria are the cell power plants.

Communication at the Level of the Synapse

The transmitter neuron is separated from the receptor neuron by a thin space or gap called the ***synaptic cleft***. There is no physical contact between neurons; the continuity between them is realized by the exchange of chemical neuromediators.[5]

The arrival of the electric signal releases small quantities of neuromediators (e.g., dopamine) that are enclosed in vesicles that open up on the membrane of the neuron transmitter. Then the neuromediators travel across the synaptic cleft and attach themselves to receptors on the other side of the gap.

The fixation of neuromediators on these receptors gives rise to another action potential in the neuron receptor (excitation) or blocks the electric signal (inhibition). This electric event is combined with other signals received by the neuron receptor at the level of its dendrites and cellular body, and the result is transmitted along the axon to the next destination.

These emissions of neuromediators occur a few times per second, and then the neuromediator (e.g., dopamine) is recaptured to avoid prolonged action and ensuing loss.

Memory at the Level of the Synapse

Apart the electrical information (i.e., excitation or inhibition of the receptive neuron), the fixation of the neuromediator on the receptor often induces some intracellular chemical events that leads to changes the way the receptive neuron works: this is the basis of the memory process. Two types of memory appear at the level of the synapse: short-term memory of a few hours and long-term memory that can last years. Short-term memory involves transient changes in the connections between neurons. Although there is no anatomical change, existing synapses are strengthened for a while to form a nonlasting memory. On the other hand, long-term memory involves enduring changes that result from the ***physical growth*** of new dendrites with synaptic connections. And when information is repeated, neural signals use the same paths, making them grow thicker with ***larger bushes of dendrites*** and corresponding synapses. The more some synaptic connections are used, the stronger they become, and the more information flows down the activated neural paths. These circuits become hardwired in the brain throughout education.

The process starts very early, even before we are born. Sensory information and motor movements shape the brain by establishing new circuits day after day. Axons set off in search of other neurons to establish new lines of communication that are reinforced with use. With learning and repetition, the circuits become more strongly embedded in the brain. This explains the major impact of events experienced in one's early years on the rest of one's life.

---

[5] Neuromediator is a generic term of representing neurotransmitters, neuromodulators, and hormones.

Furthermore, the structure of the neural cells themselves is modified. Most axons are coated with a fatty substance, ***myelin***, which acts as insulation. As this myelin sheath becomes thicker, electrical stimuli travel faster, and the networks become more efficient and self-sustaining. They develop into a sort of highway system made up of long axon fibers that link distant regions. The more a skill is practiced, and the highways are used, the thicker the network and the better and faster we are at performing that skill.

Therefore, ***education is about brain building***, and we can expect that a student would leave the classroom with a ***different brain*** from the one she walked in with—if she learned something.

More generally, we can say that our brain is constantly rewiring itself after memorable new experiences. Every fact we face, every idea we understand, and every one of our actions takes the form of a network of neurons and synapses. Learning and memorizing is thus simply the process of shaping our brain circuits, that is, hardwiring our neuronal networks.

The Paradox of Neuroplasticity

Neuroplasticity refers to the way the brain is modified in response to a stimulus. This is well illustrated by the ***snowy*** *hill metaphor*.

The first time we go skiing downhill on fresh snow, we can follow almost any path. The second time we go down that same slope, we likely follow a path close to the one we took before, though not exactly the same. If we continue skiing the whole day, we develop deeper and deeper tracks in the snow and eventually we get stuck in the rut we have traced. Intensive practice can cause the very structure of the brain to change.

According to this metaphor, it is relatively easy to start doing something in a new way, even if this needs some conscious attention. But it soon becomes difficult to leave this way and eliminate the wirings already created. In fact, brain plasticity does not always work out for the best. Good habits can become hardwired, but, if we develop bad ones, they will be difficult to correct. This explains why unlearning is often much harder than learning and why education should start early.

When experiencing great change in their personal life or at work, people need time to rewire their brain networks and let go of the mental frames that hold them back. And this is not what they feel like doing, because they rely on old routines and are not ready to give change sufficient attention, according to the bias of least effort.

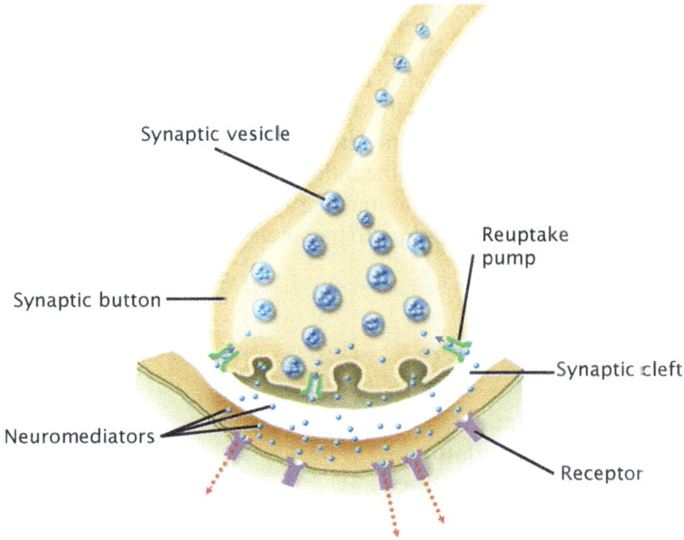

Fig. 2 Synapse

Neuromediators

Neuromediators are either *neurotransmitters* that excite or inhibit the transmission of signals, *neuromodulators* that enhance or diminish the overall effectiveness of the connection, or *hormones* that work at greater distance in the bloodstream. Some of them can have all three functions (see Fig. 2).

Neurotransmitters

For example, the neurotransmitter glutamate functions as an activator, and the neurotransmitter GABA acts as an inhibitor when neurons transmit orders from one point to another. Both act quite fast throughout the brain at the corresponding synapses.

Neuromodulators and Hormones

Neuromodulators act at the level of the extra-cellular space and regulate the transmission of signals at synapses that are already active. Their widespread action makes them useful in broadcasting the fact that something significant has happened, although it does not identify exactly what was happening. According to the American neuroscientist Joseph LeDoux, "In this, the modulatory system functions rather like an alarm going off in a small-town fire station. Whilst the alarm alerts the firemen to the fact that there is a fire, it doesn't tell them where exactly it is".

As neuromodulators or hormones modify the functioning of synapses in the nanospace of the junction, they modify our perception of reality and modulate our interior states. For example, dopamine and acetylcholine drive attention, serotonin conditions our mood, and noradrenaline influences our vigilance and alertness. This range of neuromodulators forms an amazing tool that enables us to evaluate our environment and adjust to it according to its potential interest.

The Cast of Neuromediators

Dopamine: **the stimulator**
Serotonin: **the mood regulator**
Noradrenaline: **the over-amplifier**
Acetylcholine: **the learner**
Cortisol: **the gladiator**
Testosterone: **the dominator**
Oxytocin: **the pacifier**

Annex 4—The Fear Circuit

The Amygdala[6]

The amygdala is a brain center involved in fear management (see Fig. 3). It is part of an alert system, connected with all perceptual systems (vision, hearing, touch, etc.) that constantly monitors our environment. As soon as an unforeseen event occurs, the amygdala activates and triggers the release of neuromediators and specific programs that prepare our body to flee, freeze or fight.

Through acquired experience that prepares the anticipation of what is to be expected in each situation, this sentinel estimates the emotional meaning of a new perception and controls the emotional responses. Using preprogrammed reactions associated with emotional memories, the amygdala helps regulate fear and anger and therefore plays an important role in our social interactions.

Regulating Our Emotional States and Impulses

Self-control is the ability to keep a cool head when we are confronted with a strong emotion or a threat, or when we resist reacting impulsively. It's the ***dopamine control circuit*** that is activated and a specific part of the prefrontal cortex, ***the orbitofrontal cortex, OFC***, which is situated in the lower third of the prefrontal cortex, at the bottom of the frontal lobe and just above the orbit of the eyes (hence its name). It is referred to as the ***emotional regulator***.

[6] For information, we are presenting major cerebral structures in a schematic way. They should not be considered as stand-alone modules that process a particular function (too reductive a local approach), but as structures which tend to be more particularly involved in the management of this or that function within much larger and more complex neural networks.

© The Editor(s) (if applicable) and The Author(s), under exclusive license to Springer Nature Singapore Pte Ltd. 2023

J. Teboul and P. Damier, *Neuroleadership*,
https://doi.org/10.1007/978-981-99-5122-2

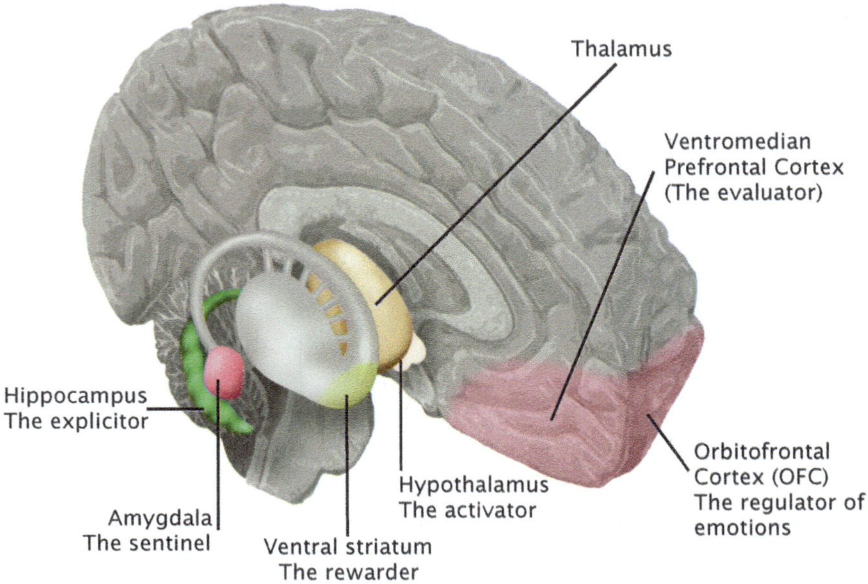

Fig. 3 Amygdala, limbic regions, and orbitofrontal cortex

The OFC therefore ***regulates emotions and desires***, helping us to slow down the emotional circuits of the limbic system. It puts a brake on the activation of deep emotional structures such as the amygdala. It also provides some cognitive flexibility by assessing the situation and allowing us to pause before acting. This voluntary braking system requires great mental energy and cognitive effort to exercise willpower and it tires quickly, like a muscle.

The inhibitory connections from the prefrontal cortex toward limbic structures (amygdala or rewarder) are much less numerous than the connections that arise from the limbic structure to the OFC. Hence, the braking system of the brain quickly reaches its limit, making us vulnerable to the appeal of emotions.

Explicit and implicit memory

We process and encode our experiences according to explicit and implicit memories.

Conscious experiences such as what we have done this week and the fact that bananas are yellow are assembled into ***explicit memories***. They are called ***explicit or declarative*** because we can explicitly refer to them during a conversation. They are mainly distributed in the internal part of the temporal cortex, where the ***hippocampus*** is located. This important structure, that we call the '***explicitor***', is the entry door for information that arouses our interest and attention.

Implicit memory is less conscious and quite vast. It begins in the womb and predominates throughout our early years. From our emotions, perceptions, actions, and bodily sensations, we create implicit mental models that shape our expectations about reality and the way we represent the world. All of this occurs without effort or intention, and our ***implicit mental models*** continue to influence our behaviors without our self-awareness.

Implicit memory is made up of two key components: procedural memory and emotional memory. ***Procedural memory*** is not directly conscious but enables us to write, walk, ride a bicycle, type text, play a musical instrument, or react spontaneously in a dangerous situation. It refers to knowledge and memories to which we do not have direct access, such as motor and perceptual skills, automatic behaviors, habits, and reflex pathways that originate in our basal ganglia (see annex 7).

The emotional memory of an experience is orchestrated by the amygdala which works closely with the hippocampus to integrate the explicit details of that experience. We remember explicitly who we were with, what we were doing, and whether the experience was good or bad. As the American neuroscientist Joseph LeDoux explains, "The hippocampus is crucial in recognizing the face you see as that of your cousin. But the amygdala adds that you don't really like her." Generally, emotionally charged incidents are remembered better than non-emotional events. For example, do you remember what you were doing on September 11, 2001? Or what you were doing during the preceding two months? It seems that our strongest emotional memories are engraved on the amygdala like on a stone tablet, and they are all the more indelible because they are implicit.

Annex 5—The Reward System

The brain measures everything we perceive using a valuation system that constantly guides our decisions. Such systems, of varying degrees of complexity exist in any animal. In fact, they assign a positive or negative value to each event or encounter to help the animal act appropriately and survive in its environment.

The Structures Involved in the Valuation Circuit[7]

In humans, the valuation circuit (Fig. 4) mainly links the ventral striatum (the rewarder), the prefrontal cortex, the amygdala (the emotional sentinel), the hippocampus (the memory explicitor), the hypothalamus (the activator), and the ventral tegmental area (the sprinkler) that sprinkles the relevant areas with dopamine.

On the biological level, a reward corresponds to the satisfaction of a need that can also result from the desire to achieve an abstract objective. When the need is fulfilled, it is often associated with a sensation of pleasure, which is amplified if the result is better than expected. The neurons of the ventral tegmental area, the sprinkler, are then activated and trigger a release of dopamine in target areas.

[7] For information, we are presenting major cerebral structures in a schematic way They should not be considered as stand-alone modules that process a particular function (too reductive a local approach), but as structures which tend to be more particularly involved in the management of this or that function within much larger and more complex neural networks.

© The Editor(s) (if applicable) and The Author(s), under exclusive license to Springer Nature Singapore Pte Ltd. 2023

J. Teboul and P. Damier, *Neuroleadership*,
https://doi.org/10.1007/978-981-99-5122-2

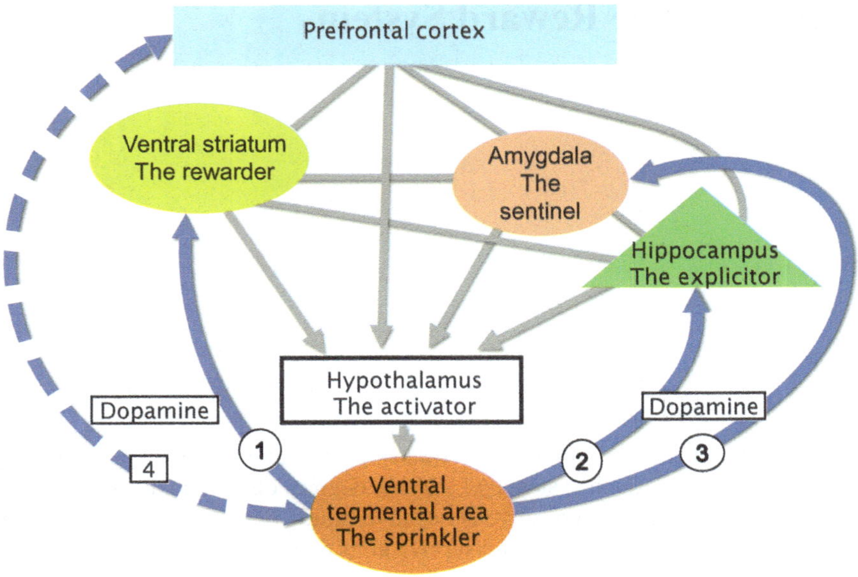

Fig. 4 Dopaminergic pathways and valuation system. 1, 2, 3 Mesolimbic pathways (desire or reward circuit) 4 Meso-cortical pathway (control circuit)

The ventral striatum, the rewarder

The ventral striatum[8] is an essential reward center that manages emotional reward, reward anticipation and motivation. It is involved in any emotion, whatever its nature. It acts as a relay, a transfer platform that controls our emotions and transmits affects to the rest of the brain. It reacts instantly and takes whatever it can take in the moment, without seeking to invest in the future. It plays a major role in learning, as it allows the stabilization of neural networks that have been activated during a successful action. This action will now be carried out with greater ease in the future. Conversely, there is no stabilization of the neuron networks activated during failure. It is at this level that trial-and-error learning is managed. Once a gesture or a behavior has been perfectly learned, it is passed onto the dorsal striatum which will control its automatic execution.

Ventral tegmental area. Reward and control circuits

The ventral tegmental area (the sprinkler) sends projections to the ventral striatum (the rewarder) and other limbic regions such as the amygdala and hippocampus. The scientific term for this ***reward or desire circuit*** is the ***mesolimbic pathway***. It is activated when we get a glass of water when thirsty, when we manage to reach a goal,

[8] The ventral striatum is also called nucleus accumbens. In rodents, the nucleus accumbens and the deep nuclei are quite well differentiated. In primates, there is more continuity between regions, so it is preferable to use the term ventral striatum.

or by proxy, when we see our favorite team win a tournament. The reward circuit drives raw desires, motivation, and enthusiasm. It is powerful, focuses attention, and points the way. It is a two-way system responding with a release of dopamine in the event of success, and a drop in the level of dopamine, if not.

The reward or desire circuit is inhibited by a complementary system, the ***control circuit***, the ***meso-cortical pathway*** that now involves the frontal cortex and its connection with the sprinkler. It manages the uncontrolled urges of the desire circuit and generates abstract thinking to evaluate options or long-term plans to dominate the world around us.

Tenacity based on the control circuit

Willpower helps us fight against desires when the reward circuit leads us too far. The control circuit regulates and dampens emotions to allow us to think with a cool head. The control circuit brings discipline and helps to anticipate further delayed reward. Life is about the future, about improvement, innovation.

Hence, there is a strong competition between the short-term orientation of the rewarder, and the long-term orientation of the prefrontal cortex and in particular a specific part of the PFC, the orbitofrontal cortex,[9] referred to as the ***emotional regulator***.

The prefrontal cortex and the orbitofrontal cortex, the OFC[10]

The OFC plays a major role in regulating and dampening emotions. It puts a brake on the activation of deep emotional structures such as the amygdala (the sentinel) and the ventral striatum (the rewarder). It also provides some cognitive flexibility by assessing the situation and allowing us to pause before acting, for example by controlling our reactions before we get angry. This voluntary braking system requires great mental energy to repel temptation. It tires quickly, like a muscle, and requires some time to recover, while deep emotional systems (limbic systems) function more automatically, without much effort.

A good metaphor for the OFC in action is to imagine it like a rider mastering his steed. He holds the reins to prevent the horse from bucking, but the force required is clearly visible and the rider may well tire before the horse. As dopamine concentration increases, neurons in the rewarder lose sensitivity to moderation signals from the OFC, the emotional regulator. Above a certain threshold, they may even stop heeding these signals. The amygdala and hippocampus then take control of the rewarder and direct behavior toward immediate rewards and pleasure. This mechanism is at the heart of dependency.

---

[9] The OFC is situated in the lower third of the PFC, at the bottom of the frontal lobe and just above the orbit of the eyes (hence its name).

[10] For information, we are presenting major cerebral structures in a schematic way. They should not be considered as stand-alone modules that process a particular function (too reductive a local approach), but as structures which tend to be more particularly involved in the management of this or that function within much larger and more complex neural networks.

The hippocampus, the memory explicitor

The hippocampus is located on the inner side of the temporal lobes. Its seahorse shape gives it its name ('hippos', meaning 'horse' and 'kampos', meaning 'sea monster'). This central structure allows us to include, store, or recall conscious, explicit memories. It is a necessary waypoint for us to remember something and put it in context.

This explicit memory is particularly sensitive to the contextual situation and can be associated with a strong emotional connotation, whether it results from a conflict or an accident, or a pleasant moment like a romantic date.

The hypothalamus, the activator

The hypothalamus is the conductor of the hormonal symphony. It controls internal states and regulates a wide range of behavioral and physiological activities, such as hunger, thirst, body temperature, or sexual activity.

Acting as a control center for the endocrine system, the hypothalamus orchestrates the release of hormones via the pituitary gland. It sends messages to the body about the emotional state of our brain in the form of hormones. They act in slower way than neurotransmitters, and in a broader range of targets.

If we are in danger, the amygdala informs the hypothalamus, which then releases cortisol and adrenaline into the bloodstream to help our body muster the resources needed for a vigorous defense. However, repeated alerts can turn into chronic stress, with a detrimental effect on the vascular and immune systems.

Annex 6—The Insula: The Body Sensor

The ***insula*** or ***insular cortex*** which is hidden under the folds of the parietal and temporal cortex that form the lateral sulcus, is the center responsible for representing and controlling our visceral functions (see Fig. 5). It informs the brain about the state of the body when there are problems or pain. By integrating a strong body reference to our emotional experiences, the insula can express 'gut' feelings such as disgust and social emotions such as contempt, guilt, or embarrassment. For this reason, we call the insula the ***body sensor***. This region is strongly linked to the ***anterior cingulate cortex***, an important structure for the detection of conflict, for example, when the result of an action is not what was expected, but also when we observe a person making contradictory statements.

These two areas also seem important for self-regulation, in noticing body sensations of pain or hunger, in recognizing one has made a mistake, or feeling the pain of being rejected or avoided. This explains why, when we have important decisions to make, it is sometimes helpful to listen to our visceral sensations—our gut feeling.

© The Editor(s) (if applicable) and The Author(s), under exclusive license to Springer Nature Singapore Pte Ltd. 2023
J. Teboul and P. Damier, *Neuroleadership*,
https://doi.org/10.1007/978-981-99-5122-2

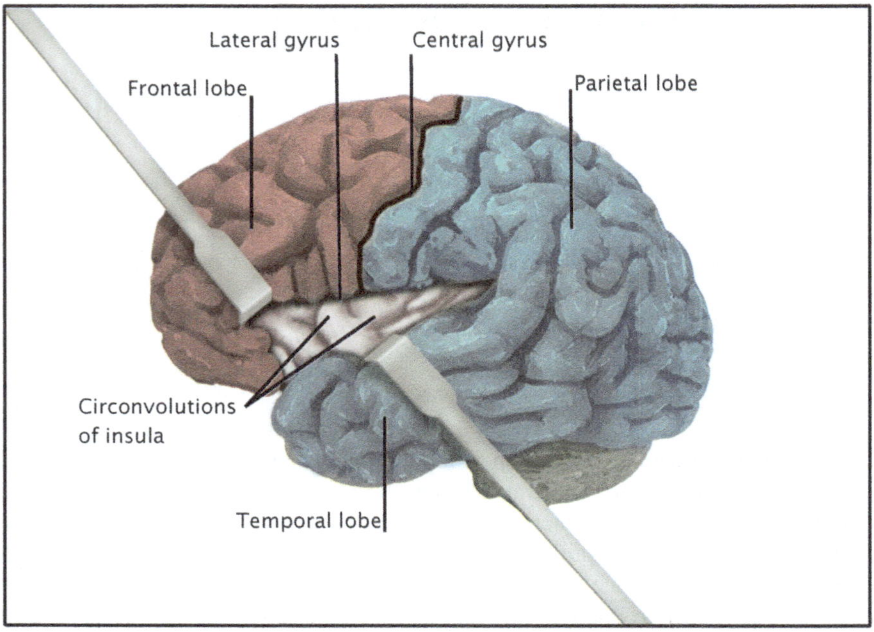

Fig. 5 Insula, the body sensor

Annex 7—Habits

Operational Automatisms

The limited capacity under conscious control of the prefrontal cortex and, more specifically of the reasoner, explains why conscious brain functioning is only a small part of the brain's whole activity, which is based for the most part on automatic procedures.

Much of what we do throughout the day, such as walking, speaking, writing, or social interaction, is automatic. Such activities once repeated thousands of times become automated and then implicit, as we don't need to think about them. They take the form of habits, which greatly ease the task of the conscious executive function. As we no longer need to deal with routines and details, we can focus our attentional energy on decision making, or thinking (Fig. 6).

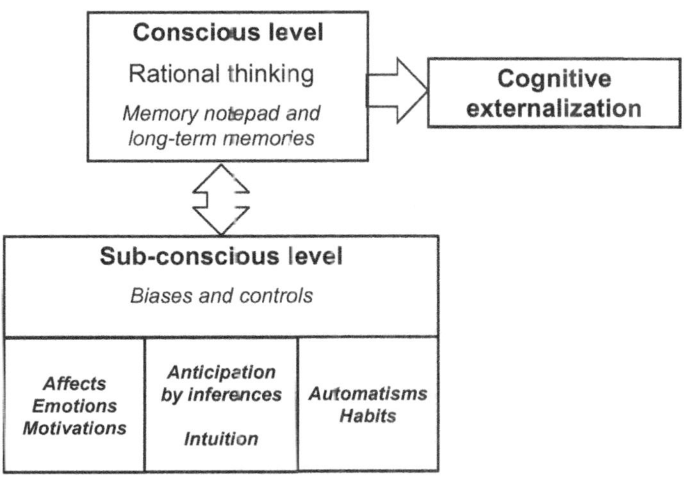

Fig. 6 Conscious and subconscious processing

© The Editor(s) (if applicable) and The Author(s), under exclusive license to Springer Nature Singapore Pte Ltd. 2023

J. Teboul and P. Damier, *Neuroleadership*,
https://doi.org/10.1007/978-981-99-5122-2

The first time we drive a car, our movements are awkward, and we have to concentrate our attention on anticipating every movement. We then go on to improve our driving by practicing and repeating the various movements, learning from our successes and mistakes. Once the skill is acquired, our autopilot comes online, and we can drive for hours while chatting with a passenger, listening to music, or thinking about something else, without consciously engaging with the act of driving. We avoid crashing into other cars, we stop at traffic lights, all without really thinking about it and paying much attention to it. These routines are performed automatically because they have been learned and over-learned. However, when faced with an unforeseen situation, such as a narrowing of the roadway, the reasoner immediately regains conscious control of the driving.

The Curators

Our automatic routines and habits are managed by neural networks that involve the ***basal ganglia***. They comprise a cluster of small structures the size of two symmetrical golf balls situated in the deep center of the brain at the base of the two hemispheres (Fig. 7). These structures appeared in the evolution of vertebrates a long time before the cerebral cortex. In humans, they operate in close connection with the cortex and require less energy to operate than conscious thinking. We call them the ***curators*** as they perform a similar function to that of a museum curator. They methodically select and filter information from the cortex and keep only the most relevant behaviors, i.e., those that provided a result in accordance with our expectations, and a release of dopamine. This is the learning phase of selecting the "works" to keep in the museum, to use our metaphor. Once acquired, this learning is added to the vast repertoire of automatic behaviors, our behavioral memory, responsible for our habits and our skills. The curators will in turn select and further adjust the learned automatic behaviors, in real time, in collaboration with the cerebellum, while using cortex information from external and internal environments, as well as the motivations of the subject.

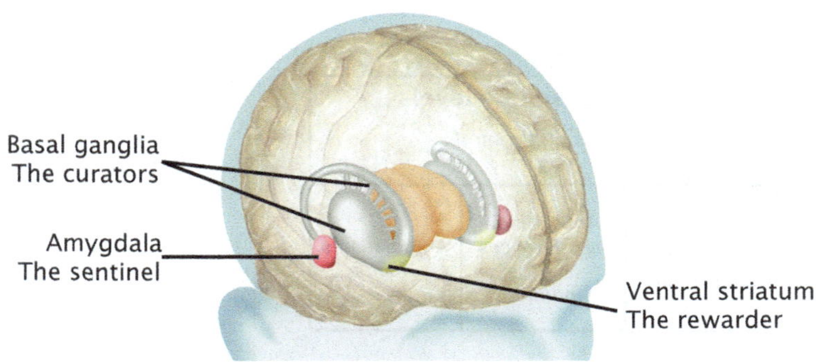

Fig. 7 Central gray nuclei or basal ganglia

The curators thus ensure harmonious actions, which are adapted to the situation and the goals of the individual. This is the phase of restitution and executive control of the learned behaviors, the time to present the "works" which are preserved in the museum, to conclude our metaphor.

How Habits and Norms Become Anchored in the Brain

Mastery of a behavior is, in fact, an action program initially under conscious control which is gradually automated, buried in the unconscious and becomes implicit. Regular practice increasingly strengthens the associated neural circuitry and its thickness. Thus, a path becomes a road, and soon a highway for the most repeated, best integrated behaviors. A good illustration of the strengthening of neural circuits is the formation of habits in the army. Recruits are prepared to react very quickly in all sorts of circumstances by instilling immediate reactions through intensive training. Oft-repeated routines thus prepare soldiers to fire instinctively, avoid an ambush or communicate under enemy fire. Each order draws on these standardized behaviors that can be executed fast and without significant costs in terms of attentional energy.

Annex 8—Teaching and Practicing Biases

Learning About the Functioning of the Brain

Any training in leadership and change management should include a specific course on the functioning of the brain, on its limitations and pitfalls. Each of these explorations should be done, one after the other, in short, repeated sessions, followed by group discussion and concrete examples from daily practice. This training should extend to the entire organization from top to bottom. The essential goal is to develop a common culture of relational engagement and learning.

This training should consider the way the brain learns efficiently, which can be broken down into four closely linked steps that can be represented by the sequence: Attention, Connection, Emotion, Spacing (ACES).[11]

Attention

Attention commands and drives learning because without it there is no access to explicit and long-term memories. Hence, to learn something, we need to give it enough attention, over enough time. Cramming learning into long teaching or study sessions can rapidly overwhelm our attentional capacity. Giving more is retaining less. Information enters and leaves the mind without being grasped or connected.

[11] Davachi L., Kiefer, T., Rock, D. and Rock, L. (2010) *Learning that lasts through AGES*. NeuroLeadership Issue 3.

© The Editor(s) (if applicable) and The Author(s), under exclusive license to Springer Nature Singapore Pte Ltd. 2023

J. Teboul and P. Damier, *Neuroleadership*,
https://doi.org/10.1007/978-981-99-5122-2

Connection and Appropriation

The role of educators cannot be reduced to the transmission of knowledge. They contribute to ***building and structuring*** their students' brains by developing and reinforcing their existing neural pathways, and by embedding the new learning in these networks. As new learning tends to connect to the existing networks, it is essential to verify that these networks are well structured without weak spots. Hence the importance of systematically checking the strength and quality of the acquired knowledge with multiple tests. Testing not only assesses what we know but changes the way that knowledge is memorized and connected to the existing networks. Repetition leads to short-term memory boost, but testing and retrieving enhances long-term retention. Memorization is also strengthened by connecting as many networks as possible. Hence the benefit of learning in multiple ways, through multiple sensory channels.

Emotion and Motivation

Emotions and motivators help capture attention and amplify memorization. To be motivated, learners need to be interested in the task, its value, and the potential reward they might get from it, which, as we know, lead to the release of dopamine. The reward might also come from the sheer pleasure of discovery. The human brain is constantly in search of novelty and enjoys it. Hence the importance of organizing new activities or varying learning techniques to strengthen motivation and capture attention. Motivation can also derive from the pleasure of receiving encouragement and positive feedback, or sharing new knowledge with colleagues or other teammates.

Spacing

Spacing learning reinforces memorization and consolidates new knowledge during periods of rest or sleep.[12] Experiments by Robert Bjork have established that spaced learning is greatly superior to massed learning. Students generally forget 90% of what they learn in class within 30 days, and most of this forgetting occurs within the first few hours. However, forgetting allows prioritizing events, and relearning in a different mode or context reinforces memorization. In fact, desirable difficulties—such as spacing study sessions, testing, and interweaving rather than blocking practice sessions—enhance retrieval and optimize long-term retention.

This is contrary to general opinion. Whatever the domain, be it military drills, athletics training, workplace learning, or school teaching, we tend to believe that block learning will lead to better retention. Beyond a certain minimal amount, repeated studying does not seem to matter much.

[12] Kornell et al. (2010).

The most efficient way of teaching should be based on delivering knowledge in brief teaching sessions, followed by practice and teamwork. Knowledge is gradually acquired over time by multiple exposures involving questioning, discussing, and sharing.

Bibliography

Alderdice, J. (2021). Conflict, complexity, and cooperation. *New England Journal of Public Policy, 33*(1), Article 9.
Ariely, D. (2008). *Predictably irrational: The hidden forces that shape our decisions*. HarperCollins.
Baars, B. (1997). *In the theater of consciousness: The workspace of the mind*. Oxford University Press.
Baars, B., & Gage, N. (2013). *Fundamentals of cognitive neuroscience: A beginner's guide*. Academic Press.
Baddeley, A. (1994). The magical number seven: Still magic after all these years? *Psychological Review, 101*(2), 353–356.
Baron-Cohen, S. (2003). *The essential difference: The truth about the male and female brain*. Basic Books.
Belbin, R. M. (1993). *Team roles at work*. Butterworth-Heinemann.
Bennis, W. (1989). *On becoming a leader*. Addison-Wesley.
Bohler, S. (2019). *The human bug*. Robert Laffont Group.
Brafman, O., & Brafman, R. (2008). *Sway: The irresistible pull of irrational behavior*. Doubleday.
Bronner, G. (2018). *Démocratie des crédules (democracy of the gullible)*. Odile Jacob.
Bronner, G. (2021). *Apocalypse cognitive*. PUF.
Brooks, D. (2011). *The social animal: The hidden sources of love, character, and achievement*. Random House.
Burns, J. M. (2010). *Leadership (Harper Perennial Political Classics)* (1st ed.). Harper Perennial Modern Classics.
Burns, D. (2020). *Feeling great*. PESI Publishing & Media.
Cialdini, R. (2016). *Pre-suasion*. Random House Books.
Coutrot, T. (2018). *Free work*. Editions du Seuil.
Coyle, D. (2010). *The talent code*. Penguin Random House.
Coyle, D. (2018). *The culture code*. Penguin Random House.
Craig, A. D. (2015). *How do you feel?* Princeton University Press.
Crozier, M. (1998). *La crise de l'intelligence*. Dunod.
Crozier, M. (1970). *La Société bloquée*. Seuil.
Cyrulnik, B., & Bustany, P. (2012). *Votre cerveau n'a pas fini de vous étonner*. Albin Michel.
Damasio, A. (1994). *Descartes' error: Emotion, reason and the human brain*. Grosset/Putnam.
Damier, P. (2014). *Décider en toute connaissance de soi. Neuroscience et décision*. Odile Jacob.
de Villiers, P. (2018). *Qu'est-ce qu'un chef ?* Fayard.

© The Editor(s) (if applicable) and The Author(s), under exclusive license to Springer Nature Singapore Pte Ltd. 2023

J. Teboul and P. Damier, *Neuroleadership*,
https://doi.org/10.1007/978-981-99-5122-2

De Vulpian, A. (2003). *A l'écoute des gens ordinaires*. Dunod.
Dehaene, S. (2014). *Le code de la conscience*. Odile Jacob.
Dehaene, S. (2007). *Les neurones de la lecture*. Odile Jacob.
Diamond, J. (2019). *Upheaval: Turning points for nations in crisis*. Little, Brown and Company.
Duhigg, C. (2012). *The power of habit: Why we do what we do in life and business*. Random House.
Dunbar, R. (2004). *The human story: A new history of humanity's evolution*. Faber and Faber.
Dupuy, F. (2005). *La Fatigue des élites*. Seuil.
Dupuy, F. (2011). *Lost in management. La vie quotidienne des entreprises au XXIe siècle*. Paris, Seuil.
Dupuy, F. (2015). *La faillite de la pensée managériale*. Seuil.
Edmondson, A. C. (2012). *Teaming: How organizations learn, innovate, and compete in the knowledge economy*. Jossey-Bass Pfeiffer.
Ehrenberg, A. (2018). *La mécanique des passions : Cerveau, comportement, société*. Odile Jacob.
Feldman Barrett, L. (2017). *How emotions are made*. Pan Macmillan.
Gladwell, M. (2005). *Blink: The power of thinking without thinking*. Black Bay Books.
Goleman, D. (1995). *Emotional intelligence*. Bantam Books.
Goleman, D. (2006). *Social intelligence: The new science of human relationships*. Bantam Books.
Graziano, M. (2013). *Consciousness and social brain*. Oxford University Press.
Hamid, N., et al. (2019). Neuroimaging 'will to fight' for sacred values: An empirical case study with supporters of an Al Qaeda associate. *Royal Society Open Science, 6*, 181585.
Haidt, J. (2006). *The happiness hypothesis: Putting ancient wisdom to the test of modern science*. Arrow books.
Harari, Y. (2015). *Sapiens: A brief history of humankind*. Harper Perennial.
Heffernan, M. (2011). *Willful blindness: Why we ignore the obvious at our peril*. Walker.
Hofstede, G., Hofstede, J., & Minkov, M. (2010). *Cultures and organizations: Software of the mind*. McGraw Hill.
Houdé, O. (2019). *Apprendre à résister*. Odile Jacob.
Huston, N. (2008). *L'Espèce fabulatrice*. Actes Sud.
Hume, D. (1779). *A treatise of human nature*, vol. 3.
Hwang, V., & Horowitt, G. (2012). *The rainforest: The secret to building the next Silicon Valley*. Regenwald.
Iacoboni, M. (2009). *Mirroring people: The science of empathy and how we connect with others*. Picador.
Janis, I. (1982). *Groupthink: Psychological studies of policy decisions and fiascoes* (2nd ed.). Houghton Mifflin.
Kahneman, D. (2011). *Thinking, fast and slow*. Allen Lane Great Britain.
Kets de Vries, M. (2006). *The leadership mystique: Leading behavior in the human enterprise*. Financial Times/ Prentice Hall.
Kornell, N., Castel, A. D., Eich, T. S., & Bjork, R. A. (2010). *Spacing as the friend of both memory and induction in younger and older adults. Psychology and Aging, 25*, 498–503.
Kouzes, J., & Posner, B. (2002). *The leadership challenge* (3rd ed.). Jossey-Bass.
Lakoff, G. (2014). *Don't think of an elephant (new edition)*. Chelsea Green Publishing Co.
LeDoux, J. (1996). *The emotional brain: The mysterious underpinnings of emotional life*. Simon & Schuster.
LeDoux, J. (2015). *Anxious: Using the brain to understand and treat fear and anxiety*. Viking.
Lieberman, M. (2013). *Social: Why our brains are wired to connect*. Crown.
Little, B. (2014). *Me, myself and Us: the science of personality and the art of well being*. Public Affairs.
Lustig, R. H. (2017). *The hacking of the American mind. Zaltbommel*. Van Haren Publishing.
Marquet, D. L., & Covey, S. R. (2013). *Turn the ship around! A true story of turning followers into leaders*. Portfolio.
Mercier, H., & Sperber, D. (2017). *The enigma of reason*. Amsterdam University Press.
Orthouse, P. (2012). *Leadership: Theory and practice*. SAGE Publications.

Panksepp, J. (1998). *Affective neuroscience: The foundations of human and animal emotions*. Oxford University Press.
Pfeffer, J. (2010). *Power: Why some people have it and others don't*. Harper Business.
Pfeffer, J. (2015). *Leadership BS*. HarperCollins Publishers.
Pink, D. (2009). *Drive: The surprising truth about what motivates us*. Riverhead Books.
Ramachandran, V. (2011). *The tell-tale brain: Unlocking the mystery of human nature*. William Heinemann.
Ripoll, T. (2020). *Pourquoi croit-on? Psychologie des croyances*. Sciences Humaines Éditions.
Rizzolatti, G., & Sinigaglia, C. (2008). *Les neurones miroirs*. Odile Jacob.
Rock, D. (2009). *Your brain at work: Strategies for overcoming distraction, regaining focus, and working smarter all day long*. Harper Business.
Rosenzweig, P. (2007). *The halo effect*. Free Press.
Rosenzweig, P. (2014). *Left brain, right stuff: How leaders make winning decisions*. Profile books.
Russo, J., & Schoemaker, P. (2002). *Winning decisions: Getting it right the first time*. Currency.
Sapolsky, R. (2017). *Behave*. Random House.
Schein, E. H., & Schein, P. A. (2018). *Humble leadership: The power of relationships, openness, and trust (the humble leadership series)*. Berrett-Koehler Publishers.
Schumpeter, J. (1942). *Capitalisme, socialisme et démocratie* (p. 1990). Payot, Bibliothèque historique Payot.
Sennett, R. (2006). *The culture of the new capitalism*. Yale University Press.
Sibony, O. (2019). *Vous allez commettre une terrible erreur !* Flammarion.
Stern, S., & Cooper, C. (2017). *Myths of management: What people get wrong about being the boss (business myths)*. Kogan Page.
Sunstein, C. (2009). *Going to extremes. How like minds unite and divide*. Oxford University Press.
Surowiecki, J. (2005). *The wisdom of crowds*. Anchor Books.
Taleb, N. (2007). *The black swan: The impact of the highly improbable*. Random House.
Tavris, C., & Aronson, E. (2007). *Mistakes were made (but not by me): Why we justify foolish beliefs, bad decisions, and hurtful acts*. Harcourt.
Teboul, J. (2006). *Service is front stage*. Palgrave Macmillan.
Teboul, J., & Damier, P. (2017). *Neuroleadership : Le cerveau face à la décision et au changement*. Odile Jacob.
Teboul, J., & Damier, P. (2022). *Le mirage du leadership à l'épreuve des neurosciences*. Odile Jacob.
Tetlock, P. (2005). *Expert political judgment how good is it? How can we know?* Princeton University Press.
Thaler, R., & Sunstein, C. (2008). *Nudge: Improving decisions about health, wealth and happiness*. Yale University Press.
Westen, D. (2007). *The political brain: The role of emotion in deciding the fate of the nation*. Public Affairs.
Wilson, D. S. (2019). *This view of life*. Pantheon Books.
Wilson, T. (2002). *Strangers to ourselves: Discovering the adaptive unconscious*. Belknap Press of Harvard University Press.
Zak, P. (2017). *Trust factor: The science of creating high-performance companies*. AMACOM.

SPRINGER NATURE

GPSR Compliance

The European Union's (EU) General Product Safety Regulation (GPSR) is a set of rules that requires consumer products to be safe and our obligations to ensure this.

If you have any concerns about our products, you can contact us on ProductSafety@springernature.com

In case Publisher is established outside the EU, the EU authorized representative is:

Springer Nature Customer Service Center GmbH
Europaplatz 3
69115 Heidelberg, Germany

The manufacturer's authorised representative in the EU is Springer Nature Customer Service Centre GmbH, Europaplatz 3, 69115 Heidelberg, Germany. If you have any concerns regarding our products, please contact ProductSafety@springernature.com

Printed and bound by CPI Group (UK) Ltd, Croydon, CR0 4YY
30/09/2026
02225314-0002